Chapter 1: Introduction to AI in Biochemical Engineering
Overview of AI Technologies and Their Applications in Biochemical and Biomedical Fields

Artificial Intelligence (AI) has revolutionized numerous industries, from finance and healthcare to entertainment and manufacturing. One of the most transformative areas in recent years has been the application of AI in biochemical and biomedical fields. AI technologies, such as machine learning, deep learning, neural networks, and natural language processing, are increasingly integrated into biochemical engineering, making processes more efficient, precise, and scalable.

In biochemical engineering, AI helps analyze vast datasets generated during research and industrial production. By processing complex patterns and identifying trends, AI can optimize biochemical processes, enabling significant advancements in areas such as drug development, disease diagnosis, and metabolic engineering. AI's ability to process large amounts of biological data, including genetic, proteomic, and metabolomic information, allows for deeper insights into human biology, leading to breakthroughs in personalized medicine and sustainable chemical production.

AI applications in biomedical fields include accelerating drug discovery, optimizing treatment plans, personalizing patient care, and improving diagnostics. In biochemical engineering, AI is used to streamline the design of more efficient biochemical pathways, aid in the development of new materials and biofuels, and optimize processes related to carbon capture and metabolic processes. As AI continues to evolve, it holds the potential to transform the field by improving human health, enhancing sustainability, and creating novel biochemical solutions.

The Role of Carbon Capture and Utilization (CCU) in Sustainability

Carbon Capture and Utilization (CCU) is a critical component of efforts to mitigate climate change and achieve sustainability. The primary goal of CCU is to capture carbon dioxide (CO_2) emissions, typically from industrial processes or the atmosphere, and convert it into valuable products. These products can range from biofuels and chemicals to materials used in various industries, including construction and manufacturing.

AI plays a pivotal role in enhancing the efficiency of CCU systems. By optimizing the capture, storage, and conversion processes, AI can help reduce the cost and energy requirements of CCU technologies. Machine learning algorithms can model CO_2 capture processes, simulate various biochemical pathways, and predict the performance of different capture materials, significantly speeding up the development of effective and sustainable CCU solutions.

AI also aids in the design of efficient biotechnological systems that use microorganisms or plants to fix carbon through processes like photosynthesis or bioremediation. These systems can be optimized with AI to produce bio-based chemicals or materials, further advancing sustainability goals. Through AI-enhanced carbon capture and utilization, industries can mitigate their carbon footprints while producing valuable and sustainable products.

AI's Transformative Potential in Enhancing Biochemical Processes, Particularly in Human Health

In the field of biochemical engineering, AI has a transformative potential, particularly in improving human health outcomes. The ability of AI to analyze and interpret vast biological datasets has revolutionized the understanding of human biology. From the study of genetic variations to the analysis of cellular behaviors, AI helps uncover new insights into how diseases develop, how cells function, and how the body metabolizes various substances.

AI's role in enhancing biochemical processes extends to drug discovery, where it can accelerate the identification of potential drug candidates by analyzing millions of chemical compounds and predicting their interactions with biological targets. In personalized medicine, AI algorithms can assess individual genetic profiles and recommend the most effective treatments for patients, minimizing the risks of adverse reactions and improving therapeutic outcomes.

In addition, AI is pivotal in advancing metabolic engineering, which involves the modification of metabolic pathways within organisms (including humans) to optimize the production of desired chemicals, biofuels, or pharmaceuticals. AI can optimize these pathways by simulating the effects of genetic modifications, predicting the outcomes of metabolic interventions, and ensuring the efficient conversion of raw materials into valuable biochemical products.

The potential for AI to revolutionize human health is vast. In terms of disease prevention, AI can help identify genetic markers for diseases, predict health risks, and suggest early interventions. In medical treatments, AI can improve the effectiveness of existing therapies by predicting how a patient will respond to specific drugs based on their unique biological makeup.

Furthermore, AI plays a role in advancing our understanding of lipid metabolism. Lipids are essential molecules involved in energy storage, cell membrane formation, and signaling. AI can help identify new ways to manipulate lipid conversion processes for therapeutic purposes, such as designing treatments for metabolic diseases, optimizing energy production, and improving the synthesis of biofuels and other valuable chemicals.

Conclusion of Chapter 1:

The integration of AI in biochemical engineering offers immense possibilities for improving human health, sustainability, and environmental protection. By harnessing the power of AI to optimize biochemical processes, particularly in areas such as carbon capture, lipid conversion, and metabolic engineering, we can unlock new solutions to the world's most pressing challenges. As we continue to explore AI's transformative potential, it is clear that its applications will be critical in shaping a more sustainable, efficient, and healthier future for humanity.

In the next chapter, we will delve into the **fundamentals of carbon capture and utilization (CCU)** in humans, exploring the mechanisms behind carbon capture and how AI enhances the efficiency and scalability of these systems. We will also examine the environmental and health benefits of AI-enhanced CCU, laying the foundation for understanding its critical role in sustainability and human health.

Chapter 2: Fundamentals of Carbon Capture and Utilization (CCU) in Humans

Understanding Carbon Capture Mechanisms in Biological Systems

The natural world has developed numerous mechanisms for carbon capture, some of which play a vital role in the carbon cycle. In humans, carbon capture is not a direct process in the traditional sense like large-scale industrial carbon capture systems, but rather an integral component of metabolic processes and cellular functions.

Carbon, in the form of carbon dioxide (CO_2), is constantly produced in the human body through cellular respiration. This is the process by which cells generate energy from glucose and oxygen, producing CO_2 as a byproduct. The human body effectively manages this waste CO_2 by transporting it via the bloodstream to the lungs, where it is expelled from the body through exhalation. While this process is highly efficient for regulating CO_2 within the human body, its focus is primarily on removal rather than long-term storage or conversion of CO_2 into usable forms.

However, recent advances in biochemical engineering have revealed potential methods to enhance the natural carbon capture processes within human systems. By integrating artificial technologies like carbon capture and utilization (CCU) systems, we can explore new ways of capturing and repurposing carbon in the human body, particularly for applications like energy production, pharmaceuticals, and bioremediation.

AI's role in advancing these processes is significant, as it can enhance our understanding of how CO2 is metabolized and create systems that optimize the way our bodies interact with carbon. Furthermore, AI can assist in developing biotechnological solutions that capture excess CO2 and convert it into value-added products—such as biofuels, biodegradable plastics, or pharmaceuticals—using natural biological systems in a controlled and efficient manner.

How AI Enhances CCU by Improving Efficiency and Scalability

AI is transforming the way we approach carbon capture and utilization by enhancing both the efficiency and scalability of these systems. One of the key challenges in carbon capture has been the energy-intensive nature of capturing and storing CO2, as well as finding suitable methods to utilize it effectively. AI offers solutions to these challenges by optimizing the design, operation, and scalability of CCU technologies.

AI-driven simulations and machine learning models allow scientists and engineers to predict the performance of different carbon capture materials and processes. These models analyze large datasets, such as environmental conditions, chemical properties, and biological responses, to determine the most effective ways to capture and store CO2. Machine learning algorithms can continuously adjust the operational parameters of CCU systems, ensuring that the process remains energy-efficient and cost-effective.

In the context of biological systems, AI can optimize the carbon conversion pathways by simulating different enzymatic processes, metabolic cycles, and microbial activity. For example, microbes can be engineered to consume CO_2 and convert it into biofuels, chemicals, or even food products. AI can enhance this process by modeling genetic modifications to increase microbial efficiency, predicting the outcomes of different biological interventions, and optimizing nutrient flows.

AI also has the potential to automate the monitoring and management of large-scale CCU systems, ensuring that carbon capture processes are continuously optimized. By automating adjustments and continuously learning from system data, AI can improve the scalability of these systems, making them viable for widespread implementation in industries like agriculture, pharmaceuticals, and energy production.

The Environmental and Health Benefits of AI-Enhanced CCU Systems

AI-enhanced CCU systems offer substantial environmental and health benefits. The most obvious benefit is the reduction of CO_2 in the atmosphere, a critical factor in combating climate change. By capturing and converting CO_2 into useful products, AI-driven CCU systems could help lower the amount of greenhouse gases in the environment, thus mitigating the impact of global warming and improving air quality.

One of the primary advantages of AI in CCU systems is its ability to optimize carbon utilization in a way that reduces environmental pollution. CO_2 can be converted into valuable products such as biofuels, which not only provide an alternative to fossil fuels but also promote a circular economy. The conversion of CO_2 into chemicals, materials, or even food can significantly reduce reliance on resource-intensive industries and contribute to a more sustainable, environmentally responsible future.

From a health perspective, AI-driven CCU technologies could help address the growing need for sustainable and renewable sources of pharmaceuticals and other bio-based chemicals. By capturing CO_2 and converting it into bio-based products, AI systems could enable the production of drugs, vaccines, and other essential health-related products with a smaller environmental footprint. This is particularly important as global health needs grow, requiring increased production of medicinal and nutritional substances without further taxing the planet's resources.

Moreover, AI can optimize the process of CO2 management in human systems, reducing the risks associated with excessive CO2 buildup in the body, such as respiratory acidosis. In critical medical scenarios, AI algorithms can help design personalized solutions to address issues related to carbon accumulation in patients with respiratory diseases or metabolic disorders. By enhancing human metabolic pathways for carbon removal, AI could play a key role in advancing medical treatments for conditions like chronic obstructive pulmonary disease (COPD), cystic fibrosis, and even obesity.

The Future of AI in Carbon Capture and Utilization in Humans

As the field of AI-driven carbon capture and utilization continues to evolve, it holds immense potential for improving the sustainability of biochemical processes and human health. AI will continue to be instrumental in discovering more efficient carbon capture materials, optimizing the carbon conversion pathways in humans, and creating novel applications for CO2 utilization in medicine and pharmaceuticals. The integration of AI with human metabolic systems offers exciting possibilities for creating new treatments and interventions that benefit both the environment and public health.

In the coming decades, AI-driven systems could become commonplace in efforts to manage carbon emissions, reduce environmental pollutants, and produce sustainable resources. As these technologies evolve, they will likely become a key component of a larger, integrated approach to sustainability, helping humanity mitigate the effects of climate change while improving the health and well-being of individuals.

Conclusion of Chapter 2:

AI is positioned to significantly enhance the way we approach carbon capture and utilization, both in industrial and biological contexts. By improving the efficiency and scalability of CCU systems, AI is paving the way for a more sustainable and healthier future. Through AI's ability to model, optimize, and automate biochemical processes, it is becoming an indispensable tool in addressing some of the most critical challenges facing humanity today. As the field progresses, AI will continue to offer transformative solutions for both environmental sustainability and human health, ensuring a more resilient and balanced future for all.

In the next chapter, we will explore the **science of lipid metabolism and conversion**, examining how AI can optimize these processes for various biochemical applications, including energy production and pharmaceutical advancements.

Chapter 3: The Science of Lipid Metabolism and Conversion

Lipids as an Essential Component in Human Metabolism

Lipids, commonly known as fats, are a diverse group of compounds that play essential roles in human metabolism. These molecules are not just energy reserves but also key structural components of cell membranes, signaling molecules, and agents involved in metabolic regulation. Their importance in cellular and metabolic processes cannot be overstated, as they support numerous vital functions, including energy storage, insulation, and cellular communication.

In the human body, lipids are primarily stored in adipose tissue, where they serve as a long-term energy reservoir. The most common form of lipid stored is triglycerides, which consist of a glycerol molecule bound to three fatty acids. When the body requires energy, these triglycerides are broken down into free fatty acids and glycerol through a process called lipolysis. The fatty acids can then be oxidized to produce energy through cellular respiration, while the glycerol can be converted into glucose through gluconeogenesis.

Beyond energy storage, lipids have a critical role in cellular membrane structure. Phospholipids and cholesterol are integral components of cell membranes, providing structural integrity and facilitating the proper functioning of cell receptors and channels. These lipids enable communication between cells, which is essential for maintaining homeostasis in the body.

Moreover, lipids are involved in the synthesis of various bioactive molecules, such as hormones, vitamins, and neurotransmitters. For example, certain fatty acids are precursors to eicosanoids, which regulate inflammation and immune responses. Cholesterol is the precursor for steroid hormones, such as cortisol and sex hormones like estrogen and testosterone. These functions demonstrate the vast and indispensable roles of lipids in maintaining human health.

Understanding Lipid Conversion Processes in the Human Body

Lipid conversion is a dynamic process in the human body that involves the breakdown, synthesis, and transformation of various lipid molecules to meet the body's metabolic needs. This process is tightly regulated by enzymes, hormones, and intracellular signaling pathways, ensuring that lipids are used efficiently and appropriately depending on the body's energy demands.

One key aspect of lipid conversion is **lipid metabolism**, which can be divided into catabolic and anabolic pathways. The catabolic processes break down lipids to release energy, while anabolic processes build complex lipid molecules from simpler ones. These processes occur primarily in the liver and adipose tissue, with the main metabolic pathways being **beta-oxidation** and **lipogenesis**.

- **Beta-oxidation**: This is the process by which fatty acids are broken down into acetyl-CoA molecules in the mitochondria, which then enter the citric acid cycle to produce ATP, the primary energy currency of the cell. This process is critical during periods of fasting or prolonged physical activity, when the body relies on stored fat as an energy source.
- **Lipogenesis**: This is the process by which excess carbohydrates and proteins are converted into fatty acids and triglycerides for storage in adipose tissue. Lipogenesis occurs when the body has more energy than it needs for immediate use, and the excess nutrients are converted into fat for future energy needs.

In addition to these core metabolic processes, lipids are also converted into signaling molecules and precursors for other vital molecules. For example, the conversion of arachidonic acid (a polyunsaturated fatty acid) into eicosanoids plays a crucial role in regulating inflammation and immune responses. Additionally, fatty acids can be modified to produce ketone bodies, which serve as an alternative energy source for the brain and muscles during periods of low carbohydrate availability.

The human body also utilizes specific enzymes to manage lipid conversions. For example, lipoprotein lipase (LPL) facilitates the hydrolysis of triglycerides in lipoproteins to release free fatty acids that can be absorbed by tissues for energy use or storage. Another important enzyme, acyl-CoA synthetase, is involved in the conversion of fatty acids into acyl-CoA derivatives that are then utilized in various biosynthetic and catabolic pathways.

How Lipid Conversion Can Be Applied to Biochemical Solutions

Lipid conversion plays a crucial role in developing several biochemical solutions, such as energy production, pharmaceuticals, and sustainable chemical production. By harnessing the natural processes of lipid conversion, scientists and engineers can develop innovative solutions for addressing both human health and environmental sustainability.

1. **Energy Production**: One of the most promising applications of lipid conversion is in the production of biofuels. Lipids, particularly in the form of algae or vegetable oils, can be converted into biodiesel, a renewable energy source. The conversion process involves the transesterification of triglycerides into fatty acid methyl esters (FAME), which are used as biodiesel. The use of biofuels helps reduce dependence on fossil fuels, thereby mitigating the impact of climate change. AI plays a significant role in optimizing this process by simulating different lipid sources and refining the conditions required for optimal biodiesel production. AI-driven models can predict the yield of biodiesel from various feedstocks and optimize the production conditions, improving both efficiency and scalability.

2. **Pharmaceutical Production**: Lipid conversion also plays a vital role in the pharmaceutical industry. Lipid-derived molecules are used in the production of liposomes, which are lipid-based vesicles used for drug delivery. Liposomes can encapsulate hydrophilic and lipophilic drugs, allowing for targeted and controlled release of therapeutic agents. AI can optimize lipid formulations for drug delivery systems, ensuring that the size, charge, and stability of liposomes are ideal for specific applications.

   Additionally, lipids and their derivatives are used in the synthesis of essential compounds like steroid hormones and vitamins. AI can aid in optimizing the biochemical pathways involved in producing these molecules, improving yield, and reducing costs associated with their production.

3. **Sustainable Chemical Production**: Another key application of lipid conversion is in the development of biodegradable plastics. Lipid-derived compounds can be used as raw materials for producing bioplastics, such as polyhydroxyalkanoates (PHA), which are biodegradable and environmentally friendly alternatives to conventional plastics. By optimizing lipid conversion pathways using AI, the production of PHA and other bioplastics can be scaled up, providing a sustainable solution to the plastic waste problem.

   In addition to PHA, other lipid derivatives can be utilized in the production of various bio-based chemicals, such as surfactants, lubricants, and solvents. AI can accelerate the optimization of these processes, enabling the creation of a circular economy where waste lipids are converted into valuable chemicals, reducing environmental pollution and dependence on petrochemicals.

4. **Metabolic Engineering**: Finally, lipid conversion can be used in metabolic engineering to create organisms that produce high-value compounds. By modifying the metabolic pathways of microorganisms such as bacteria, yeast, or algae, scientists can engineer these organisms to convert lipids into useful products like biofuels, pharmaceuticals, and specialty chemicals. AI can assist in designing these metabolic pathways, identifying key genes to manipulate, and predicting the effects of genetic modifications.

Conclusion

Lipid conversion is a fundamental process in human metabolism and plays an essential role in producing biofuels, pharmaceuticals, and sustainable chemicals. As the world seeks more sustainable solutions to address the growing energy and environmental crises, harnessing lipid conversion processes through innovative biochemical and biotechnological methods will be crucial. AI is at the forefront of these advancements, helping to optimize lipid conversion pathways and scale up these processes for industrial applications. Through AI-driven solutions, we can unlock the full potential of lipid metabolism, leading to more efficient energy production, healthier pharmaceutical treatments, and sustainable chemical alternatives.

In the next chapter, we will explore **AI's role in optimizing biochemical process pathways**, focusing on machine learning and neural networks for simulating and refining lipid conversion processes to improve efficiency and scalability.

Chapter 4: AI's Role in Biochemical Process Optimization
Overview of Machine Learning and Neural Networks in Biochemical Systems

Artificial Intelligence (AI) has become a game-changer in the optimization of biochemical processes. Among the most effective AI technologies used in biochemical engineering are **machine learning (ML)** and **neural networks**, both of which can analyze complex data, recognize patterns, and make predictions with a level of precision and speed that humans cannot match.

Machine learning refers to a subset of AI that allows systems to learn from data without being explicitly programmed. In biochemical systems, ML algorithms can be trained on vast datasets, such as those derived from laboratory experiments, clinical studies, or environmental monitoring, to identify patterns and correlations that might otherwise go unnoticed. This enables researchers to optimize biochemical processes by predicting outcomes, adjusting variables, and discovering new pathways without relying solely on traditional trial-and-error experimentation.

Neural networks, on the other hand, are a type of ML algorithm modeled after the human brain. They consist of layers of interconnected nodes (or neurons) that process information in a hierarchical manner. Neural networks are particularly adept at handling large amounts of complex, nonlinear data, making them highly useful in the biochemical field. For instance, in lipid metabolism or carbon capture processes, neural networks can predict how specific changes in molecular structures or environmental factors will affect the overall biochemical reaction.

These AI technologies work in tandem to automate, simulate, and optimize biochemical processes. By analyzing biochemical data, AI models can identify inefficiencies, propose improvements, and design new experiments that can save both time and resources. In essence, AI acts as a digital assistant, enabling scientists to move from traditional, labor-intensive methods to data-driven decision-making that enhances accuracy, scalability, and efficiency.

AI Models for Simulating and Optimizing Biochemical Pathways

AI models are particularly valuable when it comes to simulating biochemical pathways—chains of chemical reactions that occur within living organisms to maintain life. These pathways are complex and involve numerous enzymes, substrates, intermediates, and products. Traditional methods of studying these pathways can be slow and resource-intensive, as they require extensive laboratory work and manual analysis.

AI-powered simulations, on the other hand, can model these biochemical processes with great speed and accuracy. By training AI models on experimental data, researchers can simulate how different variables (e.g., concentrations of enzymes or substrates) will affect the outcome of a biochemical reaction. These models help predict optimal conditions for a given process, such as maximizing lipid conversion to biofuels or enhancing the efficiency of carbon capture in human systems.

One of the key advantages of AI simulations is that they can test thousands of variables in a fraction of the time it would take to conduct physical experiments. For example, researchers can use AI to optimize the metabolic pathways involved in lipid conversion or carbon capture, exploring various genetic modifications in microorganisms or humans to increase the yield of desired products.

In addition, AI can be used to design new biochemical pathways. By leveraging large databases of known biochemical reactions and compounds, AI models can propose novel synthetic routes that have not been previously explored. This has significant implications for industries such as pharmaceuticals, where creating new pathways for drug production or enhancing existing ones could lead to breakthroughs in treatment efficiency and cost reduction.

Real-World Examples of AI Enhancing Lipid Conversion Processes

One of the most promising applications of AI in biochemical optimization is in lipid conversion. Lipid conversion involves breaking down complex lipid molecules (e.g., triglycerides or phospholipids) into simpler molecules that can be used for energy production, drug delivery, or sustainable chemical production. This process is central to various industries, including biofuel production, pharmaceuticals, and sustainable chemicals.

AI has already demonstrated its ability to optimize lipid conversion processes, particularly in biofuel production. In biodiesel production, for example, AI models can predict the most efficient conversion of triglycerides into fatty acid methyl esters (FAMEs), a key component of biodiesel. By analyzing data from different feedstocks (such as vegetable oils or algae), AI can determine the optimal conditions for transesterification, a chemical reaction that converts lipids into biodiesel.

Similarly, in the pharmaceutical industry, AI is being used to optimize lipid conversion for drug delivery systems. Liposomes, which are lipid-based vesicles, are used to encapsulate drugs and deliver them to specific cells or tissues in the body. AI models can predict the best lipid formulations to create stable, biocompatible liposomes with ideal properties, such as size, charge, and stability. This optimization is crucial for enhancing the effectiveness of drug therapies, particularly in cancer treatment and gene therapy, where targeted drug delivery is essential.

AI has also made strides in optimizing lipid conversion for sustainable chemical production. In the production of bioplastics, for instance, AI can optimize the fermentation process of lipids to produce polyhydroxyalkanoates (PHA), a biodegradable polymer that can replace petroleum-based plastics. By simulating different fermentation conditions, AI can identify the most efficient strains of bacteria and the best conditions for PHA production, reducing costs and increasing scalability.

Moreover, AI's role in lipid conversion extends to human health applications. For instance, AI can optimize the pathways of lipid metabolism within the human body to enhance energy production or address lipid-related diseases. In cases of metabolic disorders such as obesity, diabetes, or cardiovascular diseases, AI can assist in designing personalized treatment strategies by predicting how individuals' lipid metabolism will respond to various interventions, such as diet changes or drug treatments.

Future Directions for AI in Biochemical Process Optimization

The future of AI in biochemical process optimization is incredibly promising. As AI technologies continue to evolve, they will play an even more central role in transforming how biochemical systems are studied and optimized. One key area of development is **integrating AI with genomic and proteomic data** to create more accurate models of biological systems. By combining AI with high-throughput sequencing technologies, researchers can gain deeper insights into how genetic variations impact biochemical processes, opening the door to more personalized approaches to drug development and disease prevention.

Another exciting frontier is the use of **reinforcement learning (RL)** in biochemical optimization. RL is a type of machine learning where the algorithm learns to make decisions by interacting with an environment and receiving feedback. In biochemical systems, RL could be used to optimize real-time decision-making in metabolic engineering or chemical production. For example, an RL agent could be used to continuously adjust the conditions in a bioreactor to maximize the production of a desired biochemical product, such as biofuels or pharmaceuticals, based on real-time data.

Finally, AI will increasingly be used to optimize **large-scale industrial biochemical processes**. As the world moves toward sustainable production of chemicals, biofuels, and pharmaceuticals, AI will be essential for scaling up these processes efficiently. With the ability to predict, simulate, and optimize complex biochemical reactions, AI will enable the design of systems that can produce these products at an industrial scale, reducing costs, increasing efficiency, and minimizing environmental impact.

Conclusion

AI is revolutionizing the way biochemical processes are optimized, particularly in the conversion of lipids. Through machine learning, neural networks, and advanced simulations, AI has enhanced our ability to model, predict, and optimize biochemical pathways for applications in energy production, pharmaceuticals, and sustainable chemical manufacturing. The future of AI-driven biochemical process optimization is bright, with potential breakthroughs in personalized medicine, industrial-scale production, and environmental sustainability on the horizon.

As we move forward, the integration of AI with experimental and real-world data will continue to refine our understanding of complex biochemical systems, ultimately allowing us to achieve more efficient, sustainable, and personalized solutions for human health and the environment. In the next chapter, we will explore how AI is enhancing **carbon capture within human metabolic pathways**, offering new opportunities for improving both human health and sustainability.

Chapter 5: Carbon Capture in Human Biology: A New Frontier

How Humans Naturally Capture and Utilize Carbon in Biological Systems

In biological systems, the process of carbon capture is not just about sequestering carbon dioxide (CO_2) from the atmosphere; it's intricately tied to essential metabolic processes that maintain life. While industrial methods focus on capturing CO_2 from the air or emissions and converting it into useable products, humans and other living organisms naturally manage carbon in much more complex ways.

In human biology, carbon is most prominently captured through the process of **respiration**. Every cell in the human body generates energy through cellular respiration, a series of chemical reactions that convert nutrients—primarily glucose—into energy in the form of ATP (adenosine triphosphate), with carbon dioxide being one of the byproducts. This CO_2 is carried by the blood to the lungs, where it is expelled from the body.

However, carbon's role doesn't stop at waste management; it is also a fundamental building block for the biosynthesis of macromolecules like lipids, proteins, and nucleic acids. The human body uses carbon derived from glucose and other metabolic intermediates to build these molecules, which are vital for growth, repair, and energy storage. In essence, carbon is both a waste product and a building material for the body.

Metabolic pathways like glycolysis, the citric acid cycle, and oxidative phosphorylation help to process glucose and other nutrients, converting them into energy and organic compounds. The effective management of carbon within these processes is essential for human health. Problems in these pathways can lead to metabolic disorders, such as obesity, diabetes, or cardiovascular disease.

Despite the efficiency of these natural processes, the human body does not have mechanisms to significantly store or repurpose CO_2. This is where artificial carbon capture (CCU) technologies, especially AI-driven systems, can complement and enhance the biological processes to more efficiently manage carbon in the body for better health outcomes and environmental benefits.

AI's Role in Enhancing Carbon Capture Within Human Metabolic Pathways

AI is emerging as a powerful tool to optimize and enhance natural processes, including carbon capture within the human body. By applying AI techniques such as machine learning, neural networks, and computational modeling, we can gain a better understanding of how carbon is processed and explore ways to enhance these processes in health and disease.

One way AI can enhance carbon capture is by improving **metabolic efficiency**. AI models can simulate metabolic networks in human cells, identifying key enzymes or metabolic pathways that could be modified to either increase the efficiency of carbon conversion or reduce excess CO_2 production. For example, AI could optimize **lipid metabolism**, encouraging cells to store more carbon in the form of fats or direct excess carbon toward beneficial biochemical pathways, such as the production of energy or essential compounds.

Moreover, AI can assist in understanding the interactions between carbon metabolism and other processes, such as **cellular growth** and **aging**. By analyzing large datasets from genomics, proteomics, and metabolomics, AI can identify new targets for intervention that can potentially enhance carbon utilization or reduce the body's CO_2 production in metabolic disorders or diseases such as cancer.

Additionally, AI can help develop **biological carbon capture systems** that enhance the body's ability to deal with excess carbon. For example, researchers are exploring ways to engineer microbes that could be introduced into the body to help capture CO_2 directly or convert it into more useful forms, like fatty acids or glucose. AI could optimize these microbial systems by simulating their behavior in human physiology and adjusting their genetic makeup for more efficient CO_2 capture and conversion.

AI-driven technologies may even have the potential to improve **respiratory efficiency** in humans, helping to manage excess CO_2 in cases of respiratory diseases such as chronic obstructive pulmonary disease (COPD) or cystic fibrosis. AI systems can be used to model and monitor pulmonary function, ensuring that the body effectively expels CO_2 and does not accumulate dangerous levels, especially in patients with impaired lung function.

Potential Applications for Improving Human Health and Sustainability

The combination of AI-driven carbon capture and human metabolism holds vast potential in both **healthcare** and **sustainability**. The following areas highlight how these technologies can be applied for transformative benefits:

1. **Enhanced Treatment for Metabolic Disorders**: AI can be used to design personalized interventions that optimize the body's ability to store and utilize carbon. For example, in metabolic diseases like diabetes, where the body struggles with glucose regulation, AI could optimize the body's metabolic pathways, improving insulin sensitivity or helping cells more efficiently convert excess glucose into usable energy or storage forms. AI models could simulate how changes in diet, exercise, or medication would impact carbon utilization and CO_2 production in patients, helping healthcare providers tailor treatment plans to individual needs.

2. **Personalized Carbon Management**: AI-driven metabolic engineering could enable doctors to tailor interventions based on a patient's unique metabolism. By analyzing a patient's genetic, lifestyle, and metabolic data, AI could recommend the best methods for enhancing carbon management—whether that's increasing energy production or reducing excess CO_2 in patients with respiratory issues. Personalized solutions could also help in maintaining metabolic balance, promoting healthier weight management, and managing chronic conditions related to lipid metabolism.

3. **Carbon Sequestration in Agriculture and Biotechnology**: Beyond human health, AI's potential to enhance carbon capture could extend to other biological systems, such as agriculture and biotechnology. By engineering crops or microbes to more efficiently capture and store carbon, AI could help develop sustainable agricultural practices that sequester more carbon in the soil. This could significantly reduce the overall atmospheric CO_2 levels while improving soil health and crop productivity.

4. **Sustainable Biofuel Production**: Another promising application of AI-driven carbon capture within human systems is in biofuel production. By optimizing metabolic pathways to convert excess CO_2 or carbon-based waste into biofuels, AI can enable the sustainable generation of renewable energy sources. Microorganisms, engineered with AI optimization, can be used to convert organic waste or atmospheric CO_2 into biofuels, which can then be used for energy production, reducing the reliance on fossil fuels and mitigating environmental damage.

5. **Environmental and Health Sustainability**: The integration of AI with human biology and carbon capture can also lead to large-scale improvements in sustainability. For instance, AI could assist in designing closed-loop systems where human metabolic waste (e.g., CO_2) is captured and reused to create valuable products such as biofuels, plastics, or chemicals. Such systems could significantly reduce the environmental footprint of industrial processes, aligning with global sustainability goals and offering a path toward carbon neutrality.

6. **Optimizing Carbon Storage**: One innovative application could involve AI-assisted **genetic modification** of human cells or engineered microorganisms that would improve their ability to store carbon in the form of lipids. This would not only reduce CO_2 in the bloodstream but also potentially generate valuable bioproducts. For instance, optimizing lipid storage could lead to healthier adipose tissue function or the creation of bioplastics, further promoting sustainability in healthcare and environmental industries.

Conclusion

AI's role in enhancing carbon capture within human metabolic pathways offers groundbreaking opportunities for improving both human health and environmental sustainability. By integrating AI-driven simulations and metabolic engineering, we can optimize carbon use, reduce excess CO_2 production, and enhance metabolic processes. This fusion of AI and biology represents a new frontier in personalized medicine, sustainable energy, and environmental management.

The future of AI-enhanced carbon capture in human biology holds tremendous potential. From optimizing metabolic pathways to creating systems that capture and store carbon for beneficial use, AI can transform the way we manage carbon in both individual health and global sustainability. As we continue to explore and refine these technologies, the integration of AI with human metabolism will likely become a cornerstone of future advancements in healthcare and environmental protection.

In the next chapter, we will delve into **AI-driven lipid conversion for medical and pharmaceutical applications**, exploring how AI is transforming the production of drugs, improving delivery systems, and creating new opportunities for health-related innovations.

Chapter 6: AI-Driven Lipid Conversion for Medical and Pharmaceutical Applications

The Role of Lipid Conversion in Pharmaceutical Production

Lipid conversion plays a critical role in the production of various pharmaceutical products. Lipids are essential in the formulation of drug delivery systems, as well as in the synthesis of bioactive molecules such as hormones, vitamins, and essential fatty acids. The ability to convert lipids efficiently and effectively is key to ensuring the quality and efficacy of these drugs.

Lipids serve as critical components in the formulation of **liposomal drug delivery systems**. Liposomes, which are lipid-based vesicles, are commonly used to deliver hydrophilic (water-soluble) drugs in a more bioavailable form. These vesicles can encapsulate drugs and protect them from degradation, while also enhancing their absorption into cells. This controlled release system can significantly improve the efficacy of treatments for diseases such as cancer, HIV, and autoimmune disorders.

AI's role in lipid conversion for pharmaceutical production is multifaceted. From optimizing the formulation of liposomes to improving the production processes for lipid-based drugs, AI can contribute significantly to enhancing drug delivery efficiency. AI models can predict how different lipid compositions will affect the stability, size, and release kinetics of liposomes, leading to the design of more effective drug delivery systems.

Furthermore, AI can also be used to optimize lipid conversion in the production of **hormones** such as corticosteroids and sex hormones. These lipids are critical for regulating a range of bodily functions, from metabolism to immune responses. AI can help streamline their production by optimizing metabolic pathways and identifying more efficient synthetic routes for these essential molecules.

AI in Optimizing Biochemical Processes for Drug Production and Delivery

AI technologies, especially **machine learning** and **deep learning**, have demonstrated immense potential in optimizing biochemical processes related to drug production and delivery. In traditional drug production, much of the process is empirical, relying on trial and error or expert knowledge to determine the optimal conditions for production. However, AI can enhance this process by simulating and predicting the outcomes of various production conditions, thereby improving both efficiency and scalability.

AI-driven optimization models can predict the ideal lipid formulations required for creating drug delivery systems that are both stable and effective. For instance, when designing lipid nanoparticles (LNPs) for mRNA vaccines, AI can analyze numerous factors such as lipid chain length, headgroup charge, and bilayer fluidity, allowing researchers to fine-tune the nanoparticle's size, stability, and release properties. This process, which could take weeks in traditional experimental settings, can be accelerated with AI-driven simulations, reducing both time and costs.

In the pharmaceutical industry, AI is also being leveraged to optimize large-scale production processes. AI can monitor real-time data from fermentation or enzymatic conversion processes and adjust the parameters to optimize lipid production. For example, AI systems can adjust temperature, pH, nutrient concentration, and other factors in real-time to ensure that the lipid molecules are synthesized efficiently, resulting in higher yields and less waste.

AI is also instrumental in **process analytical technology (PAT)**, which is used to monitor and control pharmaceutical manufacturing processes. AI-enhanced PAT tools can ensure that lipid conversion processes meet the necessary quality standards by providing real-time feedback and predictive maintenance, reducing downtime, and improving consistency in production.

Case Studies in AI-Driven Lipid Conversion for Health-Related Applications
Lipid-Based Vaccines and Drug Delivery Systems

lipid nanoparticles

AI models were used to simulate various lipid formulations and predict their effects on the stability and efficacy of the mRNA. By utilizing machine learning algorithms, researchers could identify the best lipid materials and production conditions to improve the stability of the mRNA, ensuring that the vaccines maintained their effectiveness at room temperature and were delivered efficiently into cells. This technology has revolutionized vaccine production, making it faster, more cost-effective, and scalable.

Personalized Lipid-Based Drug Delivery

personalized drug delivery systems

For instance, AI is being used to develop **lipid-coated drugs** that target specific tissues, such as tumor cells in cancer therapy. By adjusting the lipid composition, AI can help ensure that the drug is efficiently delivered to the tumor site while minimizing exposure to healthy tissues, thereby reducing toxicity.

Sustainable Lipid Production for Pharmaceuticals

biotechnological lipid production

For example, AI can assist in designing microbes that can convert renewable feedstocks (like agricultural waste) into valuable lipid-based pharmaceuticals. By simulating different metabolic pathways and identifying the most efficient routes, AI can enhance the yields of lipid-based compounds, leading to a more sustainable and cost-effective production process.

Looking Forward: The Future of AI-Driven Lipid Conversion in Pharmaceuticals

As AI continues to advance, its role in pharmaceutical applications will only grow. **AI-driven lipid conversion** is poised to redefine not only drug production but also how we treat a variety of diseases. Personalized medicine, where AI helps design individualized treatments based on lipid metabolism and genetic profiles, will become increasingly common, offering better outcomes and fewer side effects for patients.

Moreover, AI's ability to optimize the production of **bioactive lipids**—those involved in key biological processes such as inflammation, immune regulation, and cell signaling—could lead to the development of novel therapeutics for a variety of conditions. This includes diseases such as cancer, diabetes, cardiovascular diseases, and autoimmune disorders, where regulating lipid metabolism can be crucial for managing disease progression.

AI's power to integrate data from across disciplines—biochemistry, genomics, and clinical studies—will enable the rapid development of new lipid-based drugs and more efficient drug delivery systems. This shift towards AI-driven approaches in lipid conversion promises to lower costs, reduce waste, and improve the overall accessibility of life-saving medications.

Conclusion

The application of AI to lipid conversion for medical and pharmaceutical purposes is a revolutionary development that is reshaping the landscape of drug discovery and delivery. From optimizing lipid formulations to creating sustainable production systems, AI is unlocking new possibilities for improving human health through better, more efficient use of lipids. As the field progresses, the integration of AI will continue to drive innovation, making lipid-based therapeutics more personalized, effective, and sustainable.

In the next chapter, we will explore **how AI-driven lipid conversion processes are contributing to sustainable biochemical solutions**, with a focus on biofuels, biodegradable plastics, and addressing global environmental challenges.

Chapter 7: Sustainable Biochemical Solutions through AI and Lipid Conversion

The Importance of Sustainability in Biochemical Production

Sustainability has become one of the defining challenges of the 21st century, particularly in industries that rely heavily on natural resources and have significant environmental footprints. The biochemical sector, which includes everything from biofuels and pharmaceuticals to biodegradable plastics and industrial chemicals, is no exception. As the world's population continues to grow, so does the demand for energy, food, and medical products—leading to increased pressure on the planet's resources.

The traditional methods of biochemical production, including the extraction of fossil fuels, chemicals, and plastics, have long been criticized for their environmental impact, especially concerning carbon emissions, pollution, and waste generation. The need for **sustainable biochemical solutions** has never been more pressing. This is where AI, combined with cutting-edge biotechnology such as lipid conversion, holds transformative potential. By optimizing biological and chemical processes, AI can help create more efficient, eco-friendly alternatives to conventional production methods.

In particular, **lipid conversion**, a process where lipids (fats and oils) are transformed into various biochemical products, is gaining recognition for its ability to address environmental challenges. AI is poised to improve this process in several ways, from enhancing the efficiency of lipid conversion in biofuel production to creating biodegradable plastics, reducing reliance on petroleum-based products, and enabling waste-to-value processes.

How AI-Driven Lipid Conversion Processes Contribute to the Creation of Sustainable Biofuels and Biodegradable Plastics

The conversion of lipids into valuable products, such as **biofuels** and **biodegradable plastics**, offers a promising path toward sustainability. Lipids are abundant in both renewable biological materials (e.g., plant oils, algae, and waste animal fats) and in industrial waste streams, making them an attractive feedstock for creating sustainable products.

1. **Biofuels**

   The energy sector is one of the largest contributors to global carbon emissions. However, biofuels—fuels derived from biological materials—are seen as a viable alternative to fossil fuels. Lipid-based biofuels, such as **biodiesel**, are produced by converting triglycerides (lipids) from plant oils or animal fats into fatty acid methyl esters (FAMEs) through a chemical process known as **transesterification**. These biofuels emit far fewer greenhouse gases compared to conventional fossil fuels.

   AI contributes to the optimization of biofuel production in several ways. By simulating the behavior of various feedstocks, AI can predict which types of lipids will yield the highest energy output, allowing for better feedstock selection. Additionally, AI can optimize the transesterification process by fine-tuning variables such as temperature, pH, and catalyst concentration, ensuring that the conversion of lipids into biofuels is as efficient as possible.

2. **Biodegradable Plastics**

   Plastics are a major environmental challenge due to their persistence in the environment and their reliance on petroleum-based feedstocks. Lipid-derived **bioplastics**, such as **polyhydroxyalkanoates (PHA)**, are an environmentally friendly alternative. PHA is produced by microorganisms, such as bacteria or algae, which convert lipids into this biodegradable polymer.

   AI can significantly improve the production of PHA and other biodegradable plastics by optimizing the microbial fermentation processes involved. By simulating the metabolic pathways of microorganisms and analyzing data from large-scale fermentation experiments, AI can identify which strains of bacteria are most efficient at producing PHA and under what conditions. These optimizations can lead to higher yields of bioplastics, making them more cost-competitive with traditional plastic products.

Exploring the Potential for Using AI to Address Global Environmental Challenges

The potential applications of AI-driven lipid conversion for sustainability go beyond biofuels and biodegradable plastics. As AI continues to evolve, it will increasingly be used to address broader environmental challenges. The following are some of the key areas in which AI-powered lipid conversion can have a significant impact:

1. **Waste-to-Value Processes**

AI can optimize the conversion of waste materials—such as agricultural by-products, food waste, and municipal solid waste—into valuable biochemical products. These waste-to-value processes can help reduce environmental pollution while simultaneously creating biofuels, biodegradable plastics, and other useful chemicals. By modeling different waste streams and predicting which lipids or waste materials can be converted into the most efficient products, AI can help create a circular economy where waste is transformed into valuable resources, minimizing the environmental impact of production processes.

2. Carbon Sequestration and Carbon-Neutral Production

The development of AI-driven lipid conversion processes has the potential to make a significant contribution to global efforts to combat climate change. Lipid-based biofuels and bioplastics offer the possibility of **carbon-neutral production**, where the carbon emitted during the production and use of these bio-based products is offset by the carbon absorbed by the plants or microorganisms used to produce the lipids in the first place. By optimizing carbon capture and lipid conversion, AI can help ensure that these processes are as carbon-efficient as possible, playing a vital role in reducing overall greenhouse gas emissions. Additionally, AI can assist in the development of other carbon-sequestration technologies, such as algae-based carbon capture systems, which absorb CO_2 from the atmosphere and convert it into lipids or other useful products. These systems could become a key component of strategies aimed at achieving carbon neutrality and addressing the global climate crisis.

3. **Sustainable Agriculture and Food Production**

   AI-driven lipid conversion processes can also contribute to sustainable agriculture by optimizing the production of bio-based chemicals and biofuels from crops and waste. By improving the yield of crops that are rich in lipids (such as algae or oilseeds), AI can help ensure that the agricultural sector contributes to both food security and the production of sustainable fuels and materials. Furthermore, AI can assist in reducing agricultural waste by identifying which parts of the crop can be converted into biofuels or other useful chemicals, promoting a more efficient and environmentally friendly use of resources.

4. **Reducing Plastic Pollution**

   By optimizing the production of biodegradable plastics using lipid conversion processes, AI can help address the global plastic pollution crisis. Traditional plastics are derived from petrochemicals, which are non-renewable and take hundreds of years to degrade. In contrast, lipid-based bioplastics, like PHA, break down naturally in the environment. AI's ability to enhance the scalability of PHA production could make it a viable alternative to conventional plastics, reducing the burden of plastic waste on the environment.

AI's Role in Advancing Sustainable Biochemical Solutions

The role of AI in advancing sustainable biochemical solutions cannot be overstated. AI technologies—such as **machine learning**, **deep learning**, and **genetic algorithms**—can rapidly analyze complex datasets, optimize biochemical processes, and predict the outcomes of various experimental conditions. In lipid conversion, AI is already enabling innovations in the production of biofuels, bioplastics, and other valuable chemicals from renewable resources.

Moreover, AI's ability to integrate data from across disciplines—biochemistry, genomics, environmental science, and engineering—will continue to push the boundaries of sustainable production. The increasing sophistication of AI models will allow for **holistic optimization**, ensuring that not only the yield of desired products is maximized, but that the environmental and economic impacts are minimized.

By improving the efficiency of lipid conversion processes, AI will help reduce the dependency on fossil fuels, decrease greenhouse gas emissions, and lower the environmental footprint of industrial production. These advancements align with the growing global push for more sustainable practices in every sector, from energy to manufacturing, and from agriculture to healthcare.

Conclusion

AI-driven lipid conversion technologies are at the forefront of sustainable biochemical solutions. Through optimizing processes like biofuel and biodegradable plastic production, AI is helping create a more sustainable future by reducing reliance on non-renewable resources, improving waste management, and contributing to carbon neutrality. As AI continues to advance, its potential to address pressing global environmental challenges will only grow, making it an indispensable tool for building a more sustainable and resilient world.

In the next chapter, we will explore **AI-optimized biofuels and renewable energy production**, focusing on how AI applications are advancing renewable energy sources and contributing to global sustainability goals.

Chapter 8: AI-Optimized Biofuels and Renewable Energy Production

The Role of Lipids in Biofuel Production

Biofuels have emerged as one of the most promising alternatives to fossil fuels, offering a sustainable solution to the growing energy demands of the world while reducing the environmental impact associated with carbon emissions. **Lipids**, particularly in the form of **vegetable oils, animal fats, and algae**, play a crucial role in the production of biofuels. Lipids are primarily used to produce **biodiesel**, which is a renewable energy source made through the chemical conversion of triglycerides (lipids) into fatty acid methyl esters (FAMEs).

The importance of lipids in biofuel production lies in their high energy content. Triglycerides, the storage form of fats in plants and animals, are composed of three fatty acids attached to a glycerol backbone. When these fats are broken down, the fatty acids can be chemically converted into biodiesel. This process, called **transesterification**, replaces the glycerol with methyl or ethyl alcohol, producing biodiesel and glycerin as byproducts. Biodiesel has several advantages over traditional petroleum diesel, including lower emissions of greenhouse gases, greater biodegradability, and reduced toxicity.

However, the efficiency of lipid conversion into biofuels depends heavily on various factors such as feedstock quality, reaction conditions, and catalyst selection. **AI-driven optimization** is increasingly playing a central role in enhancing biodiesel production by analyzing data from diverse sources, simulating different conversion processes, and automating production adjustments to achieve maximum efficiency.

AI Applications in Optimizing Lipid Conversion to Biofuels

AI is transforming the way biofuels, specifically biodiesel, are produced from lipids by optimizing every aspect of the process, from feedstock selection to production conditions. The following are some of the ways in which AI is contributing to the optimization of biofuel production:

1. **Feedstock Selection and Quality Control**

The first step in producing biodiesel is selecting an appropriate feedstock, which can vary from vegetable oils (soybean, canola, palm) to waste oils (used cooking oils) and even algae. AI can help identify the best feedstocks for biodiesel production by analyzing factors like lipid content, cost, availability, and environmental impact. Through machine learning models, AI can predict the yield of biodiesel from different feedstocks and recommend the most efficient source. For example, AI can predict how different types of algae, which have high lipid content, can be cultivated in various environmental conditions, enabling the selection of algae strains that will provide the highest lipid yield. AI can also monitor the quality of feedstocks in real-time, ensuring that the oil is free from contaminants, which can affect the efficiency of the transesterification process.

2. Optimizing Reaction Conditions

AI plays a critical role in optimizing the **transesterification process**, which involves converting triglycerides into biodiesel. AI-driven simulations and machine learning algorithms can model how different reaction conditions—such as temperature, pressure, catalyst concentration, and time—affect the yield of biodiesel. By adjusting these variables in real-time, AI can maximize biodiesel production, reduce energy consumption, and minimize waste.

Advanced AI models, including **neural networks** and **genetic algorithms**, can be used to search for the optimal reaction conditions by analyzing large datasets from experiments and historical production runs. This data-driven approach not only improves the efficiency of biofuel production but also makes the process more economically viable by reducing operating costs.

3. **Improving Process Control and Automation**

   AI is increasingly being used to automate and control large-scale biofuel production systems. Through **process analytical technology (PAT)**, AI can monitor and control various aspects of the production process, such as mixing, heating, and separation. By continuously analyzing real-time data from sensors, AI can make adjustments to the process, ensuring that the biodiesel production is efficient and consistent.

   For example, AI-powered systems can detect variations in temperature or chemical concentrations and make instantaneous adjustments to maintain the desired reaction conditions. This not only improves the yield but also enhances the overall sustainability of the process by reducing the need for manual intervention and optimizing resource use.

4. **Predictive Maintenance and Process Optimization**

AI's ability to predict when equipment will fail or when maintenance is required is another key factor in optimizing biofuel production. Using **predictive analytics**, AI can analyze sensor data to forecast when mechanical components such as pumps, reactors, or filters are likely to fail. This allows for **proactive maintenance**, reducing downtime and ensuring the continuous operation of biofuel production plants.

In addition to maintenance, AI can also predict when changes in the production process might lead to inefficiencies. By analyzing historical data, AI can forecast production bottlenecks or areas where the yield of biodiesel could be improved, allowing operators to make adjustments before these issues impact production.

5. **Enhancing the Sustainability of Biofuel Production**

   AI can also contribute to the **sustainability** of biofuel production by optimizing the entire supply chain, from feedstock procurement to final product distribution. By simulating various production scenarios and supply chain logistics, AI can minimize the environmental impact of biofuel production, reducing waste, energy use, and carbon emissions.

   AI-driven solutions can also assist in improving the **carbon footprint** of biodiesel production. For example, AI can help select feedstocks that not only maximize biodiesel yield but also have a lower environmental impact. Additionally, AI can help optimize transportation routes for raw materials and finished biofuels, reducing fuel consumption and emissions associated with transportation.

Case Studies of AI's Role in Advancing Renewable Energy Sources and Sustainability

1. **Algae-Based Biofuel Production**

   One promising area where AI is making an impact in biofuel production is in the development of **algae-based biofuels**. Algae are rich in lipids, which can be converted into biofuels. However, algae cultivation is still a costly and energy-intensive process. AI is being used to optimize algae growth conditions, such as light exposure, nutrient concentration, and temperature, to maximize lipid yield. Researchers are using AI models to analyze environmental data and genetic information from different algae strains to identify those that have the highest lipid content and can be grown in the most efficient conditions. For instance, AI algorithms have been used to optimize the conditions for the cultivation of **microalgae** in photobioreactors, which are used to produce algal biomass for biofuel production.

   Additionally, AI is being applied to monitor the algae cultivation process in real-time, enabling researchers to identify issues such as nutrient deficiencies or temperature fluctuations that can reduce lipid production. AI-driven process control can ensure that optimal conditions are maintained throughout the cultivation period.

2. **Waste Cooking Oil to Biodiesel**

   AI is also helping improve the process of converting waste cooking oils, a major waste product, into biodiesel. Waste oils are often underutilized, but by using AI to optimize the transesterification process, it is possible to convert this waste into high-quality biodiesel efficiently. AI can optimize the selection of catalysts and reaction conditions, making the process more cost-effective and reducing the environmental impact of waste disposal.

   Several pilot projects are using AI-driven systems to monitor and control biodiesel production from waste oils, demonstrating how AI can make biofuel production from waste more scalable and economically viable. By optimizing every step of the process, AI ensures that waste oils are converted efficiently, while minimizing the consumption of additional resources like fresh feedstocks.

Conclusion

AI-optimized biofuel production represents a significant advancement in the quest for renewable energy sources. By improving the efficiency of lipid conversion into biofuels, AI is helping to make biofuels a more competitive alternative to fossil fuels. Whether through optimizing feedstock selection, improving reaction conditions, automating production processes, or reducing the carbon footprint, AI plays a crucial role in advancing sustainable energy solutions.

The future of renewable energy lies in leveraging AI to optimize not only biofuel production but also the entire lifecycle of energy generation, from cultivation and extraction to transportation and consumption. As AI continues to evolve, its role in advancing **sustainable biofuel production** will only grow, contributing to a more sustainable and carbon-neutral energy future.

In the next chapter, we will explore **metabolic engineering and AI-driven lipid conversion**, focusing on how AI is used to optimize metabolic pathways in both humans and microorganisms, enhancing lipid conversion for industrial, medical, and pharmaceutical applications.

Chapter 9: Metabolic Engineering and AI-Driven Lipid Conversion
The Concept of Metabolic Engineering in Humans and Microorganisms

Metabolic engineering is a field that focuses on the modification of metabolic pathways in living organisms to enhance the production of desired products. This process involves altering the flow of metabolites through a cell's biochemical network to increase the production of a particular compound. In humans, metabolic engineering can be used to optimize energy production, improve health outcomes, and treat metabolic diseases. In microorganisms, metabolic engineering has been used extensively to increase the production of biofuels, pharmaceuticals, and other valuable chemicals.

The foundation of metabolic engineering lies in the understanding of **metabolic pathways**, which are networks of chemical reactions that occur within cells to maintain life. These pathways involve a series of enzymes that convert substrates into intermediate molecules and ultimately into final products. By understanding how metabolites flow through these pathways, scientists can identify key enzymes or steps that can be modified to increase or decrease the production of specific compounds.

In the context of lipid conversion, metabolic engineering has become particularly important as it allows researchers to optimize lipid biosynthesis pathways for the production of biofuels, bio-based chemicals, and pharmaceuticals. For example, the conversion of algae or plant oils into biodiesel relies on the optimization of lipid metabolism to maximize the yield of fatty acid methyl esters (FAMEs), the key components of biodiesel.

AI-driven metabolic engineering takes this process to the next level by using computational models, machine learning algorithms, and data analysis to accelerate the design, optimization, and implementation of metabolic changes. AI's ability to handle large datasets and predict how different genetic modifications will affect metabolic pathways is transforming the speed and accuracy of metabolic engineering.

How AI Assists in Engineering Metabolic Pathways to Optimize Lipid Conversion

AI is revolutionizing metabolic engineering by enabling the optimization of complex metabolic pathways in microorganisms and human cells. Through **machine learning** and **systems biology**, AI can identify the most efficient metabolic routes for converting lipids into useful products. Here are some key ways AI is assisting in lipid conversion optimization:

1. **Pathway Prediction and Design**

   AI can simulate and predict the flow of metabolites through metabolic networks. By analyzing existing biochemical data, AI models can identify potential bottlenecks in lipid biosynthesis or lipid degradation pathways, providing insights into which enzymes or genes need to be modified to optimize lipid production. These AI models can also suggest entirely new pathways that may not have been previously considered, potentially opening new avenues for lipid conversion.

   For example, in the case of biofuel production from lipids, AI can predict the most efficient metabolic routes for the conversion of fatty acids into biofuels such as biodiesel. By modifying microbial genomes, AI can optimize the metabolic processes in microorganisms to increase lipid yield, enhance the efficiency of transesterification reactions, and minimize byproduct formation.

2. **High-Throughput Screening for Genetic Modifications**

In metabolic engineering, genetic modifications are made to microorganisms or human cells to enhance their metabolic pathways. AI can help identify which genes should be targeted for modification and predict the effects of these modifications on lipid metabolism. Machine learning algorithms can analyze vast datasets from high-throughput screening experiments, where different genetic modifications are tested to see how they impact lipid conversion.

Through **genomic databases** and **CRISPR-based gene editing**, AI can quickly identify candidate genes for modification, enabling faster development of genetically engineered strains that produce higher yields of lipids or lipids for specific uses, such as drug production or biofuels.

3. **Optimizing Bioreactor Conditions for Lipid Conversion**

Once microorganisms or human cells have been engineered to optimize lipid conversion, they are often cultured in bioreactors under specific conditions. AI can be used to monitor and optimize these bioreactor conditions, such as temperature, pH, oxygen levels, and nutrient concentrations, to maximize lipid production. AI-driven **process control systems** can adjust parameters in real time based on sensor data, ensuring that the system operates at peak efficiency.

By using **feedback loops** and **predictive modeling**, AI can forecast how changes in environmental conditions will impact the lipid production process, helping researchers and manufacturers fine-tune their bioreactors to achieve the best possible results.

4. **Synthetic Biology and AI-Enhanced Pathway Construction**

   AI is playing an increasingly important role in **synthetic biology**, which involves the design and construction of new biological parts, devices, and systems. By combining AI with synthetic biology techniques, researchers can design entirely new pathways for lipid conversion that do not occur naturally. AI can optimize these synthetic pathways by modeling how new genes, enzymes, and metabolites will interact within the cell.

   For instance, synthetic biology can be used to design microorganisms that convert carbon dioxide (CO_2) or waste biomass into lipids, which can then be converted into biofuels or other valuable chemicals. AI can model how the introduction of new genes into a microorganism's genome will affect lipid production and guide the selection of the best synthetic routes for these engineered systems.

Applications in Industrial, Medical, and Pharmaceutical Sectors

AI-driven metabolic engineering is being applied across a wide range of industries, from **industrial biofuel production** to **medical applications**. The following are key sectors where AI and metabolic engineering are transforming lipid conversion processes:

1. **Biofuel and Bio-Based Chemical Production**

    In the energy sector, AI-optimized lipid conversion is enhancing the production of biofuels such as biodiesel, bioethanol, and biogas. By engineering microorganisms to produce high yields of lipids from renewable sources, such as algae or agricultural waste, AI is making biofuel production more cost-effective and sustainable. AI models can also optimize the production of bio-based chemicals from lipids, reducing dependence on fossil fuels and promoting a circular economy.

2. **Pharmaceutical Production**

    Lipid conversion is critical in the pharmaceutical industry, particularly for producing drug delivery systems and bioactive lipids. By using AI to optimize the production of **liposomes** (lipid-based drug carriers) and other lipid-derived drugs, the pharmaceutical industry can enhance the efficiency of drug delivery, improve patient outcomes, and reduce costs. AI can also optimize the production of essential lipids such as **prostaglandins**, which are used in a variety of therapeutic applications, including anti-inflammatory drugs and pain management.

3. **Medical Biotechnology and Treatment of Metabolic Disorders**

   AI-driven metabolic engineering is also advancing the development of medical treatments for **metabolic disorders** such as obesity, diabetes, and cardiovascular diseases. By optimizing lipid metabolism in human cells, AI can help develop more effective treatments for these conditions. For example, AI can be used to identify new drug targets or metabolic pathways that can be modulated to regulate lipid storage or breakdown, leading to improved treatments for conditions such as non-alcoholic fatty liver disease (NAFLD) or atherosclerosis.

4. **Personalized Medicine**

   AI is increasingly being used in personalized medicine, where lipid metabolism is tailored to individual patients based on their genetic makeup. By analyzing a patient's genetic profile and understanding how their body processes lipids, AI can help design customized treatments for metabolic conditions. This could involve optimizing lipid conversion pathways in human cells to reduce the risk of diseases related to lipid metabolism, or developing more effective drug delivery systems using patient-specific lipid formulations.

Conclusion

AI-driven **metabolic engineering** and **lipid conversion optimization** represent a revolutionary approach to improving biochemical processes across a variety of industries. By harnessing the power of AI to simulate metabolic pathways, optimize bioreactor conditions, and enhance synthetic biology techniques, researchers are able to improve the production of biofuels, pharmaceuticals, and other valuable chemicals. These advancements hold the potential to significantly reduce the reliance on fossil fuels, increase sustainability in manufacturing, and offer new treatments for metabolic diseases.

As AI continues to evolve, its ability to accelerate the pace of metabolic engineering and lipid conversion will only increase, contributing to more sustainable and efficient production systems, better healthcare solutions, and a cleaner environment. In the next chapter, we will explore the **ethical considerations** surrounding AI-driven biochemical processes, focusing on the need for responsible development and regulation in the field of metabolic engineering and lipid conversion.

Chapter 10: Ethical Considerations in AI-Driven Biochemical Processes

Addressing the Ethical Implications of AI in Biomedical and Biochemical Fields

The integration of Artificial Intelligence (AI) into the biomedical and biochemical fields holds transformative potential, from improving human health outcomes to creating sustainable bio-based products. However, as with any powerful technology, the use of AI in these domains raises important ethical questions that must be addressed to ensure that AI's impact is both responsible and beneficial to society. The ethical considerations surrounding AI-driven biochemical processes span several areas, including privacy, equity, safety, accountability, and the potential for unintended consequences.

One of the primary ethical concerns in AI-driven biochemical engineering is ensuring that **health-related applications of AI** are aligned with the best interests of patients and the general public. This includes ensuring that AI systems are used to improve health outcomes in a way that is **fair, transparent**, and **non-discriminatory**. AI algorithms, when used to optimize drug production, lipid conversion, or personalized treatments, must be developed with a commitment to fairness, ensuring that these systems do not inadvertently reinforce existing health disparities based on race, socioeconomic status, or geographic location.

Another significant concern is **patient privacy**. As AI systems increasingly rely on large datasets, including personal health information, it is essential that this data be handled with the utmost care. Strict regulations must be in place to protect patient privacy and ensure that data is used ethically and securely. Furthermore, **informed consent** is vital. Patients must be fully informed about how their data will be used, especially when participating in AI-driven clinical trials or studies.

AI bias is also an ethical challenge in the biochemical and biomedical fields. If the data used to train AI models is not representative of diverse populations, the resulting AI systems could perpetuate biases. For example, if an AI model is trained predominantly on data from one ethnic group, it may not perform as well for individuals from other ethnicities, leading to health disparities. The potential for biased outcomes necessitates diverse and inclusive data collection and rigorous testing of AI systems to ensure fairness across all demographics.

Ensuring Responsible AI Development in Human-Related Biochemical Applications

To ensure responsible AI development in biochemical applications, several principles and frameworks must be followed:

1. **Transparency and Explainability**

   One of the key requirements for responsible AI development is transparency. AI models, especially in biomedical and biochemical contexts, must be understandable to both experts and non-experts. While AI systems may be inherently complex, it is important to make their decision-making processes as transparent as possible. This is particularly critical when AI is used to make healthcare decisions, such as recommending treatments or optimizing lipid conversion processes. Stakeholders, including doctors, patients, and regulators, should be able to understand how AI arrived at a particular conclusion or recommendation.

   In this context, **explainability** is a key factor. AI systems in healthcare must not operate as "black boxes." Instead, the rationale behind AI-driven decisions must be communicated clearly, ensuring that medical professionals can trust and verify the outcomes.

2. Accountability and Regulation

As AI becomes more integrated into biochemical and biomedical systems, it is crucial that there is a clear system of **accountability**. This includes assigning responsibility for the outcomes of AI-driven decisions, especially when it comes to patient safety. If an AI system makes an error or causes harm, it is important to determine who is responsible—whether it be the developers of the AI, the healthcare providers, or another party.

Regulatory bodies must also adapt to keep pace with the rapid advancements in AI technologies. Governments and regulatory agencies must implement robust frameworks that ensure AI applications in biomedicine comply with **ethical standards, patient safety guidelines**, and **international best practices**. These regulations should focus on ensuring the safety, security, and efficacy of AI systems, as well as maintaining public trust in these technologies.

3. **Human Oversight and Control**

Despite the promise of AI in optimizing biochemical processes and enhancing healthcare, **human oversight** remains essential. AI should be viewed as a tool to assist, not replace, healthcare professionals and scientists. **Human-in-the-loop** systems, where AI outputs are always reviewed and validated by experts, are critical in maintaining control over decisions that impact human lives. For instance, while AI can optimize lipid conversion pathways for biofuel production or drug delivery, final decisions in medical contexts should be made by qualified healthcare professionals.

In addition, it is important to maintain human autonomy and prevent AI from making decisions that could have far-reaching consequences without human involvement. AI should be viewed as an assistant, not an autonomous decision-maker, in health and biochemical systems.

Ethical Boundaries and Regulation in AI-Driven Lipid Conversion

The ethical concerns related to AI-driven lipid conversion—whether in biofuel production, pharmaceuticals, or health-related applications—are significant. When AI is used to enhance lipid conversion processes, it must be done in a way that prioritizes sustainability, human well-being, and equity. For example, in the production of **biofuels from lipids**, AI should ensure that feedstocks are sourced responsibly, without contributing to deforestation or other environmental harm. The benefits of AI-driven lipid conversion must not come at the expense of the environment or vulnerable populations.

Regulation is also critical when AI is used to modify biological systems, particularly when AI-driven processes are being scaled up in industrial settings. For example, **genetically modified organisms (GMOs)** used in lipid conversion must be carefully regulated to prevent unintended ecological consequences. This includes ensuring that modified microorganisms used for biofuel or biodegradable plastic production do not escape into the environment, where they could potentially disrupt local ecosystems.

Another key concern is the **sustainability** of AI-driven biochemical solutions. AI can optimize lipid conversion processes to produce biofuels, biodegradable plastics, and other chemicals, but the raw materials for these processes must be sourced sustainably. For instance, if the feedstock for biodiesel production comes from crops, AI should help ensure that these crops are grown in a way that does not lead to deforestation or other forms of environmental degradation. Sustainable practices should be embedded into the very design of AI-driven biochemical systems.

Conclusion

AI's potential to enhance biochemical and biomedical applications is vast, offering the ability to optimize lipid conversion processes, improve human health outcomes, and address global sustainability challenges. However, with this power comes a responsibility to ensure that these technologies are used ethically. From ensuring **transparency** and **accountability** to promoting **human oversight** and **sustainable practices**, ethical considerations must guide the development and application of AI in these fields.

As we continue to harness AI's potential, it is crucial to maintain a careful balance between innovation and responsibility, ensuring that these technologies are used for the collective good. Moving forward, **regulatory frameworks** and **ethical guidelines** will play a pivotal role in shaping the future of AI-driven biochemical processes, ensuring that these innovations benefit society as a whole while safeguarding human rights, environmental integrity, and public trust.

In the next chapter, we will explore the role of AI in **personalized medicine and lipid conversion**, focusing on how AI is transforming individualized healthcare treatments and improving outcomes through tailored lipid metabolism solutions.

Chapter 11: AI in Personalized Medicine and Lipid Conversion
How AI Can Personalize Lipid Metabolism and Treatment for Individual Patients

Personalized medicine, often referred to as **precision medicine**, is an emerging field that tailors medical treatment to the individual characteristics of each patient. By incorporating genetic, environmental, and lifestyle data, personalized medicine enables healthcare providers to offer treatments that are more effective and have fewer side effects. The application of AI in personalized medicine has great potential, particularly in the area of **lipid metabolism**. Since lipids play crucial roles in energy storage, cellular function, and signaling, understanding how each individual metabolizes lipids is key to addressing a wide range of diseases, including metabolic disorders, cardiovascular diseases, and obesity.

AI can analyze vast amounts of data from genetic, genomic, and proteomic studies to better understand lipid metabolism in individuals. These AI models can predict how a person's body will process different lipids, helping to design personalized treatment plans. For example, AI can identify patients who may be at higher risk for lipid-related disorders due to genetic variations in enzymes responsible for lipid metabolism. This data allows for tailored interventions, such as specific dietary recommendations, personalized pharmaceutical treatments, or adjustments in lipid metabolism pathways, based on the individual's genetic makeup and metabolic profile.

Moreover, AI-driven models can simulate how changes in lipid metabolism, whether through diet, drugs, or genetic modifications, will impact an individual's health. By continuously monitoring a patient's response to treatment, AI can help adjust therapy in real-time, improving outcomes and minimizing adverse effects.

Applications of AI in Creating Personalized Pharmaceutical Solutions

AI is also revolutionizing the development of **personalized pharmaceutical solutions**, particularly in drug design and delivery systems. Traditional drug development approaches often follow a "one-size-fits-all" model, but this does not take into account the genetic and metabolic differences among individuals. Personalized medicine, aided by AI, moves beyond this model by tailoring drugs to suit an individual's unique biochemistry.

1. **AI in Drug Discovery**

   AI can enhance the drug discovery process by analyzing patient data, including genetic, proteomic, and lipidomic profiles, to identify which compounds will be most effective for treating specific lipid-related disorders. Through **machine learning** algorithms, AI can analyze millions of potential drug candidates and predict which molecules are most likely to interact with targeted lipid pathways, offering more targeted therapies. This process not only speeds up drug discovery but also reduces the costs and risks traditionally associated with the development of new drugs.

2. **Liposome-Based Drug Delivery Systems**

   Lipid-based drug delivery systems, such as **liposomes**, are commonly used to encapsulate drugs, enhancing their stability and controlled release. AI can be used to optimize the size, charge, and composition of liposomes for specific patients, based on their individual lipid metabolism profiles. This could lead to more effective and less invasive drug delivery, ensuring that drugs are delivered directly to the target tissues, such as tumors, while minimizing side effects on healthy tissues.

3. **AI in Nutritional Recommendations for Metabolic Disorders**

AI can analyze an individual's lipid metabolism to create personalized dietary plans that optimize lipid profiles and reduce the risk of metabolic disorders. For example, AI-driven systems can analyze how different types of fats (saturated, unsaturated, trans fats) affect lipid profiles in specific patients. Based on this data, AI can recommend personalized diets designed to manage conditions like **hyperlipidemia**, **obesity**, and **diabetes**—disorders where lipid imbalances are often at the core. By integrating AI with data from wearable devices and continuous glucose monitors, personalized dietary interventions can be continuously adjusted for maximum benefit.

Real-World Case Studies of AI in Personalized Medicine

1. **AI-Driven Lipid Profiling for Heart Disease**

   One promising application of AI in personalized medicine is in **cardiovascular disease**. Lipid imbalances, such as elevated levels of low-density lipoprotein (LDL) cholesterol, are major contributors to heart disease. AI models, utilizing **lipidomic profiling**, have been used to predict how different patients will respond to various lipid-lowering drugs like statins. By analyzing genetic data and lipid levels, AI can predict which medications will be most effective for each individual, optimizing treatment strategies and improving patient outcomes.

   For instance, researchers have applied AI to analyze data from clinical trials, identifying genetic markers that indicate how patients metabolize cholesterol-lowering drugs. This information allows for more targeted treatments that reduce the risk of heart disease, with fewer side effects compared to generalized treatments.

2. **AI in Diabetes Management**

AI is also playing a role in the personalized treatment of **diabetes**, particularly in understanding how lipids affect insulin resistance. Lipid metabolism has a direct impact on insulin sensitivity, and AI-driven models are being used to predict how a patient's lipid profile influences their likelihood of developing Type 2 diabetes. By integrating AI with wearable health devices like continuous glucose monitors, doctors can tailor insulin therapy to an individual's metabolic needs in real time. AI can also be used to create personalized **lifestyle interventions**, including diet and exercise programs, that help manage lipid levels and insulin sensitivity. In combination with genetic information, AI can help predict a patient's future risk of diabetes and recommend preventative actions.

3. **AI-Optimized Cancer Treatments**

   Another exciting application of AI in personalized medicine is its use in developing **lipid-based drug delivery systems** for cancer treatments. Liposomes and other lipid-based nanoparticles are used to deliver chemotherapy drugs more effectively to cancer cells, reducing the side effects that often accompany conventional chemotherapy. AI models can analyze how lipids interact with different types of cancer cells and optimize the size and composition of lipid-based nanoparticles for individual patients.

   Researchers have used AI to design liposomal formulations that target specific cancer types based on lipid metabolism. For example, AI models can help determine the best lipid composition for targeting tumors in the liver, pancreas, or lungs, optimizing the drug delivery process and improving treatment outcomes.

Enhancing Human Health Through AI-Optimized Lipid Conversion

AI's role in personalized medicine extends beyond drug discovery and delivery to broader areas of health optimization, including **lipid metabolism disorders**. AI can assist in diagnosing and managing diseases that involve lipid imbalances, such as **non-alcoholic fatty liver disease (NAFLD)**, **hypercholesterolemia**, and **metabolic syndrome**. These conditions are often linked to disrupted lipid metabolism, making early detection and personalized treatment crucial.

1. **Early Detection of Lipid-Related Diseases**

   One of the most promising applications of AI in personalized medicine is its ability to detect lipid-related diseases in their early stages. AI systems can analyze a patient's lipid profile—along with genetic, lifestyle, and environmental factors—to identify patterns that indicate an increased risk for conditions like cardiovascular disease or liver disease. Early detection allows for more timely interventions, reducing the risk of long-term complications.

2. **Predicting Lipid-Related Diseases**

   AI models can be used to predict the likelihood of lipid-related diseases in individuals based on their lipid metabolism and genetic predisposition. By analyzing vast datasets from electronic health records, wearable devices, and genomic studies, AI can identify individuals who are at risk for diseases such as atherosclerosis, diabetes, or obesity, allowing for preventive measures or personalized treatment plans.

3. **Optimizing Treatment for Metabolic Disorders**

   For patients already diagnosed with lipid-related disorders, AI can help optimize treatments. By continuously monitoring the patient's lipid levels and other health indicators, AI can provide recommendations for lifestyle changes or medication adjustments. Over time, AI systems can learn from the patient's response to treatment, fine-tuning the approach for the best possible health outcome.

Conclusion

AI's application in personalized medicine and lipid conversion is revolutionizing healthcare by offering treatments that are tailored to the individual needs of each patient. From optimizing lipid metabolism to designing personalized drug delivery systems, AI is helping to ensure that patients receive the most effective treatments with the fewest side effects. Through **genomic analysis, predictive models**, and **real-time monitoring**, AI is creating a future where healthcare is truly personalized, improving both the quality and outcomes of medical treatments.

As AI technology continues to evolve, its potential to revolutionize medicine—especially in the realm of lipid metabolism—will only grow. In the next chapter, we will explore **enhancing human health through AI-optimized lipid conversion**, particularly in the treatment of metabolic disorders and the optimization of biochemical processes for better health outcomes.

Chapter 12: Enhancing Human Health Through AI-Optimized Lipid Conversion

Understanding Lipid-Related Diseases and Disorders

Lipids, which include fats, oils, cholesterol, and phospholipids, are crucial components of all biological cells and perform a variety of essential functions in the human body. They are involved in storing energy, forming cell membranes, and regulating various signaling pathways that influence processes like immune response, metabolism, and cell growth. However, when lipid metabolism becomes imbalanced, it can lead to a range of **lipid-related diseases**.

Common lipid-related disorders include **obesity, hyperlipidemia, cardiovascular diseases, non-alcoholic fatty liver disease (NAFLD)**, and **diabetes**. These conditions often arise when the body is unable to properly process or store lipids, either due to genetic factors, lifestyle choices, or environmental influences. For instance, obesity and cardiovascular diseases are frequently associated with an excess of **low-density lipoprotein (LDL)** cholesterol in the bloodstream, which can lead to plaque buildup in arteries, increasing the risk of heart attacks and strokes.

Similarly, **insulin resistance**—often linked to excess body fat—can disrupt lipid metabolism, contributing to the development of Type 2 diabetes. NAFLD, another major concern, is characterized by the accumulation of fat in the liver, which can progress to more severe liver conditions like cirrhosis or liver cancer if left untreated.

Understanding these diseases and their connection to lipid metabolism is critical for developing effective treatments and interventions. By leveraging **artificial intelligence (AI)**, we can accelerate the discovery of solutions to treat these disorders, improve patient outcomes, and create personalized treatments that are more effective and safer.

How AI-Driven Lipid Conversion Can Contribute to Treatments for Metabolic Conditions

AI's role in lipid conversion is emerging as a game-changer in the treatment of metabolic conditions and other lipid-related diseases. By optimizing how the body processes lipids and fine-tuning biochemical pathways, AI can play an essential role in both preventing and managing these disorders.

1. **Optimizing Lipid Metabolism**

   AI-driven models can simulate the lipid metabolic network within the human body, identifying disruptions or inefficiencies in the process. By analyzing vast datasets from genetic studies, proteomics, and metabolomics, AI can pinpoint the molecular causes of lipid-related diseases. For example, AI can identify **genetic variations** in enzymes like **lipoprotein lipase (LPL)** or **acetyl-CoA carboxylase**—key regulators in lipid metabolism—that contribute to conditions such as hyperlipidemia or obesity.

   Once these pathways are understood, AI can help develop drugs or therapies that **restore normal lipid metabolism**, either by enhancing enzyme activity, suppressing harmful lipids, or promoting the synthesis of beneficial lipids. AI can also identify potential interventions that normalize **cholesterol levels**, reduce **fatty acid accumulation** in the liver, or restore **insulin sensitivity**, addressing the root causes of diseases like diabetes and cardiovascular conditions.

2. **Personalized Treatment for Obesity and Metabolic Syndrome**

One of the major challenges in treating obesity and metabolic syndrome is the variability in how individuals respond to diet, exercise, and medications. AI can offer a more personalized approach by analyzing an individual's **genetic profile**, **lipid metabolism**, and **lifestyle factors** to predict how they will respond to specific treatments. By integrating data from wearable devices, AI can monitor a person's lipid levels, blood sugar, and body composition in real time, adjusting recommendations for **nutrition** and **exercise** accordingly.

For example, AI can analyze a person's response to dietary interventions—such as **low-fat, low-carbohydrate**, or **ketogenic diets**—and predict the best approach for reducing **visceral fat** or improving **insulin sensitivity**. This allows for treatments that are tailored to the unique metabolic needs of each individual, improving the effectiveness of interventions and reducing the risk of side effects.

3. **Drug Development for Lipid-Related Disorders**

In the pharmaceutical industry, AI is being used to design and optimize **lipid-targeting drugs**. AI algorithms can analyze large amounts of data from clinical trials, genetic studies, and biochemical assays to identify compounds that interact with lipid metabolism pathways in beneficial ways. For example, **PCSK9 inhibitors**, a class of drugs used to lower LDL cholesterol, were discovered using AI-driven approaches that analyzed the genetic basis of cholesterol regulation. This is just one example of how AI can accelerate the development of novel lipid-targeting therapeutics.

Additionally, AI is being used to optimize **lipid-based drug delivery systems**, such as liposomes and lipid nanoparticles, which can deliver drugs more effectively to specific tissues. AI-driven design can ensure that these delivery systems are **biocompatible**, stable, and able to release the drug at the desired rate, improving treatment outcomes for patients with lipid-related diseases.

4. **AI in Lipid-Driven Cancer Therapies**

   Cancer therapies, particularly those targeting **tumor metabolism**, are increasingly focusing on lipid metabolism. Many cancers alter their lipid profiles to support rapid growth and survival. AI can help identify which lipid metabolism pathways are upregulated in specific types of cancers and suggest targeted treatments to block these pathways.

   Lipid-based drug delivery systems, such as liposomal chemotherapy drugs, can be optimized using AI to ensure that these treatments specifically target tumors while minimizing side effects to healthy tissues. AI can model the behavior of lipid-based drug carriers and predict how different formulations will perform in delivering chemotherapeutic agents to cancer cells. This ensures that cancer treatments are more effective and less toxic, offering improved quality of life for patients.

The Potential for AI to Improve Human Health Outcomes Through Optimized Biochemical Processes

AI's ability to model, optimize, and simulate lipid conversion processes has profound implications for improving human health. By enhancing lipid metabolism and developing personalized treatments, AI can help address some of the most pressing metabolic diseases, such as **cardiovascular disease**, **diabetes**, **obesity**, and **fatty liver disease**.

1. **Early Detection and Prevention of Lipid-Related Diseases**

   AI is being used to enhance the **early detection** of lipid-related diseases. By analyzing a patient's lipid profile in conjunction with other data points—such as genetic markers and lifestyle factors—AI can predict an individual's risk of developing diseases like heart disease or diabetes before symptoms appear. This allows for early interventions that can prevent or delay the onset of these conditions.

   AI can also be used to **predict** how a patient's condition will progress, providing healthcare providers with the information needed to adjust treatment plans over time. For example, AI could identify patients who are at risk of developing fatty liver disease or cardiovascular complications based on their lipid metabolism and suggest personalized treatments or lifestyle changes to mitigate these risks.

2. **Optimizing Biochemical Pathways to Maximize Health**

   AI is enabling the **optimization of biochemical pathways** related to lipid metabolism to enhance overall health. By continuously monitoring lipid levels, glucose metabolism, and other biomarkers, AI systems can optimize metabolic processes to promote healthy aging, increase energy levels, and improve overall cellular function. This includes optimizing **fatty acid oxidation** pathways to support weight management or boosting **lipid signaling** to improve immune system function.

3. **Improved Management of Chronic Conditions**

    For patients with chronic lipid-related conditions, AI can assist in managing their health more effectively. By integrating real-time data from wearable devices, AI can offer personalized treatment recommendations for managing conditions like diabetes, obesity, and hyperlipidemia. For example, AI can suggest modifications in diet, exercise, or medication to help patients achieve better lipid balance, optimize insulin sensitivity, and reduce the risk of complications.

4. **Enhanced Personalized Wellness**

    AI can also play a role in improving **personalized wellness** by analyzing how different people metabolize lipids, carbohydrates, and other macronutrients. Personalized wellness programs, developed using AI, could recommend lifestyle interventions that optimize lipid metabolism and overall health. For example, AI could suggest the optimal macronutrient ratio for individuals based on their genetic profile, lifestyle, and current metabolic state, promoting healthier lipid levels and metabolic function.

Conclusion

AI-driven lipid conversion processes are helping to revolutionize the treatment of lipid-related diseases, offering a more personalized, efficient, and sustainable approach to healthcare. By optimizing lipid metabolism pathways, AI is playing a critical role in the prevention, diagnosis, and treatment of conditions such as **obesity**, **cardiovascular disease**, and **diabetes**. As AI continues to advance, its ability to improve human health outcomes through optimized biochemical processes will only increase, providing new opportunities for healthier lives and more effective treatments.

In the next chapter, we will explore the integration of **AI and genetic engineering** to enhance lipid conversion pathways, focusing on the potential of AI-driven gene editing and synthetic biology in advancing health and sustainability.

Chapter 13: AI in Genetic Engineering and Lipid Conversion

The Integration of AI and Genetic Engineering to Enhance Lipid Conversion Pathways

In recent years, **genetic engineering** and **artificial intelligence (AI)** have emerged as powerful tools for optimizing biological systems, especially in the context of **lipid conversion**. These two fields, when integrated, have the potential to revolutionize the way we approach metabolic processes, creating more efficient and sustainable pathways for the production of biofuels, pharmaceuticals, and other valuable biochemicals.

Genetic engineering involves the direct manipulation of an organism's DNA to alter its biological functions. In microorganisms such as bacteria or yeast, genetic engineering can be used to enhance lipid production by introducing or modifying genes that control lipid metabolism. On the other hand, AI is utilized to model, predict, and optimize these genetic changes, making it possible to rapidly scale up processes and predict the outcomes of genetic modifications in complex biological systems.

The combination of AI and genetic engineering allows for a more precise, data-driven approach to modifying lipid metabolism in humans and microorganisms. AI tools can analyze vast amounts of data from genomic, transcriptomic, and metabolomic studies to uncover how specific genes and metabolic pathways influence lipid production and conversion. This enables researchers to make informed decisions about which genetic modifications are likely to yield the most promising results.

How Gene Editing Can Complement AI-Driven Processes in Metabolic and Lipid-Based Solutions

Gene editing technologies such as **CRISPR-Cas9, TALENs (Transcription Activator-Like Effector Nucleases)**, and **Zinc Finger Nucleases** have revolutionized the field of genetic engineering, making it possible to make precise modifications to an organism's DNA. These tools allow researchers to knock out, activate, or replace specific genes within a microorganism's genome, enabling them to optimize lipid production or alter lipid metabolism in highly specific ways.

AI complements these gene-editing tools by predicting the effects of these genetic modifications on lipid metabolism. For example, AI models can simulate the impact of knocking out a gene involved in lipid storage or activating a gene that promotes fatty acid biosynthesis. Once a promising genetic modification is identified, CRISPR or other gene-editing technologies can be used to make the desired change in the organism's DNA, resulting in a microorganism that is optimized for lipid conversion.

Moreover, AI can assist in **high-throughput screening** to test large libraries of genetic modifications, identifying the most effective alterations for enhancing lipid metabolism. By integrating AI with gene editing, researchers can rapidly iterate through different genetic strategies, ultimately identifying the best candidates for industrial applications such as biofuel production or pharmaceutical synthesis.

The Future of AI in Synthetic Biology and Genetic Modification

Synthetic biology is an interdisciplinary field that combines principles from biology, engineering, and computer science to design and construct new biological parts, devices, and systems. By combining **AI** with **synthetic biology**, scientists can create entirely new metabolic pathways in organisms that do not naturally exist, enabling the production of novel compounds and biochemicals, including lipids and lipophilic drugs.

AI's ability to process vast amounts of genomic data enables the design of **synthetic genetic circuits** that can regulate lipid metabolism in microorganisms. These engineered circuits can be fine-tuned to respond to environmental signals or other stimuli, ensuring that lipid production is optimized in real-time. For example, AI-driven models can design circuits that automatically adjust lipid production in response to changes in nutrient availability or temperature, improving yield and consistency.

One of the most exciting potential applications of AI in synthetic biology is the development of **microbial factories** that produce biofuels, pharmaceuticals, or other valuable compounds from simple feedstocks, such as carbon dioxide, waste biomass, or agricultural by-products. By engineering microorganisms to efficiently convert these raw materials into lipids or other bio-based products, we can create a more sustainable, circular economy, where waste is turned into valuable resources.

AI's Role in Optimizing Lipid Production for Industrial and Pharmaceutical Sectors

In both **industrial biotechnology** and the **pharmaceutical industry**, the optimization of lipid conversion processes is a key focus area. Microorganisms, such as bacteria, yeast, and algae, are being engineered to produce lipids at scale for a wide range of applications. These lipids are not only used as biofuels but also in the production of **biodegradable plastics, food additives, cosmetic ingredients**, and **pharmaceuticals**.

1. **AI-Driven Biochemical Pathway Optimization for Biofuel Production**

The production of biofuels, such as **biodiesel**, relies on efficient lipid conversion pathways. Through genetic engineering, microorganisms can be modified to produce higher yields of lipids that are then converted into biodiesel. AI plays a crucial role in **optimizing metabolic pathways** for maximum lipid production, ensuring that microorganisms grow efficiently, utilize feedstocks effectively, and produce high-quality biofuels. By combining genetic engineering with AI, scientists can also ensure that these processes are scalable, cost-effective, and sustainable, meeting the growing demand for renewable energy sources.

2. **Pharmaceutical Applications of Lipid-Based Drug Delivery**

Lipid-based systems, such as **liposomes**, are widely used in the pharmaceutical industry to deliver drugs more efficiently to targeted tissues, minimizing side effects. AI can enhance the design of these systems by predicting the optimal lipid compositions that will improve the bioavailability of drugs. By simulating how different lipids interact with drug molecules, AI can guide the design of lipid carriers that increase the stability, release rate, and targeting accuracy of pharmaceutical compounds.

In addition, AI can optimize the production of **lipid-soluble drugs**, such as certain vitamins, hormones, and anticancer agents, by identifying the best genetic modifications to promote lipid synthesis in microbial systems. This allows for the large-scale, cost-effective production of these drugs, which is particularly important in addressing global health challenges.

The Role of AI in Synthetic Biology and Personalized Health Solutions

AI's role in **synthetic biology** extends beyond industrial applications. In **personalized medicine**, AI can be used to tailor lipid-based therapies to individual patients. By analyzing genomic data, AI can predict how a patient's body will respond to certain lipid-modulating treatments, such as statins or lipid-based drug delivery systems. This allows healthcare providers to create customized treatment plans that are more likely to succeed, reducing the trial-and-error approach that often accompanies drug therapy.

Moreover, AI can assist in the development of **gene therapies** that alter a patient's lipid metabolism at the genetic level. For example, gene editing techniques like CRISPR can be used to correct genetic mutations that lead to inherited lipid disorders, such as familial hypercholesterolemia. AI-driven models can identify the most effective gene-editing strategies, predict potential off-target effects, and ensure that the treatment is both safe and effective.

Challenges in AI-Driven Genetic Engineering for Lipid Conversion

Despite its immense potential, the integration of AI and genetic engineering to enhance lipid conversion is not without its challenges. Some of the key obstacles include:

1. **Ethical and Regulatory Concerns**

   The use of genetic engineering and AI in human health raises ethical questions, particularly in relation to gene editing and synthetic biology. For example, **germline editing** (modifying the DNA of embryos or reproductive cells) has raised concerns about the potential for unintended consequences or misuse. Ethical guidelines and regulatory frameworks must be in place to ensure that genetic engineering is used responsibly and for the benefit of all.

2. **Complexity of Biological Systems**

   Biological systems are incredibly complex, and the interactions between genes, proteins, and metabolites are not fully understood. While AI can help predict and optimize genetic changes, it is not always possible to foresee every outcome, particularly in large-scale industrial applications. Further research and collaboration between AI experts, genetic engineers, and biologists will be essential in overcoming these challenges.

Conclusion

The integration of **AI and genetic engineering** is reshaping the future of lipid conversion and its applications across a range of industries. By optimizing lipid metabolism pathways through genetic modifications and AI-driven models, we are able to create more efficient, sustainable, and personalized solutions in biofuel production, drug development, and medical treatments. While challenges remain, the potential for these technologies to revolutionize health, sustainability, and industry is immense.

In the next chapter, we will explore the **role of AI in environmental sustainability through lipid conversion**, focusing on how AI can help reduce carbon emissions, promote green chemistry, and facilitate the conversion of waste into valuable bio-based chemicals.

Chapter 14: AI in Environmental Sustainability through Lipid Conversion

How AI Optimizes Carbon Capture and Lipid Conversion to Support Environmental Sustainability

As the world grapples with climate change and environmental degradation, finding innovative solutions to reduce carbon emissions and promote sustainability has never been more urgent. One such solution lies in the optimization of **carbon capture** and **lipid conversion**, processes that can play a vital role in mitigating the environmental impact of human activities. AI technologies have proven to be powerful tools in advancing these processes, offering new approaches to address the pressing challenges of carbon neutrality and sustainable production.

Carbon capture and utilization (CCU) refers to the process of capturing carbon dioxide (CO_2) emissions from the atmosphere or industrial sources and converting them into valuable products, such as biofuels, biodegradable plastics, or other biochemicals. Lipids, which are crucial molecules in both energy storage and metabolic processes, can serve as effective carbon sinks. By enhancing lipid conversion pathways, we can not only improve carbon capture but also produce sustainable products that contribute to a circular economy.

AI plays a pivotal role in **optimizing these processes** by enabling better control, faster simulations, and more efficient scaling. By modeling the interactions between carbon dioxide, microorganisms, and lipid biosynthesis pathways, AI systems can predict and guide the most effective strategies for carbon capture, optimizing the conditions under which CO_2 is converted into valuable lipids or other biochemicals.

AI's ability to analyze massive datasets allows for real-time optimization of carbon capture processes, improving the efficiency and scalability of lipid conversion systems. AI-driven models can simulate various carbon capture scenarios, evaluate the effectiveness of different lipid-producing strains, and suggest optimal conditions for maximizing carbon fixation while minimizing energy consumption.

The Impact of AI in Reducing Carbon Emissions and Promoting Green Chemistry

AI is uniquely positioned to drive significant progress in the reduction of **carbon emissions** by optimizing both **carbon capture** and the conversion of captured carbon into valuable products. The adoption of AI-driven solutions can lead to more efficient and cost-effective carbon capture systems, helping industries meet carbon reduction targets while simultaneously creating sustainable alternatives to fossil fuels.

1. **Enhancing Carbon Capture Efficiency**

   AI can help improve the efficiency of traditional carbon capture methods, such as **direct air capture (DAC), amine scrubbing**, and **carbon capture from power plants**. By using AI to analyze sensor data from carbon capture systems, industries can optimize operational parameters like temperature, pressure, and flow rate. This allows for more precise control of the capture process, ensuring that carbon emissions are removed from the atmosphere at a lower cost and with higher efficiency.

2. **Carbon Fixation via Lipid Conversion**

AI can also assist in the conversion of CO_2 into **lipids** through engineered microorganisms. For example, algae, bacteria, and yeast can be genetically modified to enhance their lipid production capabilities. AI can help identify the best metabolic pathways to optimize CO_2 fixation and convert it into lipids that can then be utilized for biofuel production or other biochemicals. By simulating different genetic modifications and environmental conditions, AI can identify the most efficient strains for carbon sequestration, speeding up the development of **sustainable biofuels**.

Algae are one of the most promising candidates for CO_2 fixation because of their ability to photosynthesize and convert carbon dioxide into lipids. AI can optimize algae cultivation conditions, such as light intensity, nutrient levels, and temperature, to maximize lipid production. AI can also predict the optimal genetic modifications that would enhance algae's natural ability to absorb CO_2 and convert it into valuable lipids.

3. **Promoting Green Chemistry**

   Green chemistry is the design of chemical processes that minimize the use of harmful substances, reduce waste, and increase energy efficiency. AI can assist in advancing green chemistry by optimizing lipid conversion processes, enabling the sustainable production of biochemicals and biofuels. AI-driven models can identify chemical pathways that minimize energy input, reduce carbon emissions, and generate valuable byproducts, such as biodegradable plastics and bioplastics, that replace petroleum-based alternatives.

   For example, AI can help design processes for the **conversion of waste biomass** into lipids, which can then be used as a source of biofuels or other bio-based chemicals. This **waste-to-value** process not only helps reduce waste but also minimizes the need for fossil fuel-derived raw materials.

AI's Role in Waste-to-Value Processes: Converting Waste into Useful Bio-Based Chemicals

One of the key benefits of AI in **environmental sustainability** is its ability to optimize **waste-to-value** processes. These processes involve converting waste materials, such as agricultural residues, food waste, or CO_2 emissions, into valuable products like biofuels, biodegradable plastics, and other sustainable chemicals. AI plays an essential role in identifying the most efficient pathways for converting waste into usable lipids or other bio-based chemicals.

1. **Biomass-to-Biofuels Conversion**

   AI can be used to enhance the conversion of **biomass** (organic waste) into biofuels. Biomass is a rich source of lipids and other compounds that can be used for energy production. AI models can simulate various processes, such as **fermentation, hydrolysis**, and **transesterification**, to predict how different types of biomass can be converted into biofuels. These models can help optimize the use of enzymes, microbes, and catalysts to maximize lipid yields and improve the overall efficiency of the biomass-to-biofuel conversion process.

   Additionally, AI can assist in designing more efficient **bioreactors** for biomass conversion, ensuring that the right conditions are maintained for maximum lipid production. By continuously monitoring environmental factors such as temperature, pH, and nutrient levels, AI systems can dynamically adjust these parameters to optimize the conversion process in real time.

2. **Waste CO$_2$ Conversion into Value-Added Products**

   AI can be used to guide the conversion of **CO$_2$ emissions** into valuable biochemicals such as lipids, which can then be used in biofuel production, plastics, or pharmaceuticals. Microorganisms engineered to fix CO$_2$ and convert it into lipids can be optimized using AI to enhance their ability to perform this task. By modeling various genetic modifications and environmental conditions, AI can identify the best strategies to improve CO$_2$ fixation and lipid production in these microorganisms.

   Furthermore, AI can help design bioreactors and process control systems that efficiently capture CO$_2$ from industrial emissions and direct it to microorganisms for conversion. These AI-driven systems can reduce the environmental impact of industrial activities by turning waste CO$_2$ into valuable products, promoting a circular economy.

The Role of AI in Achieving Sustainable Carbon Neutrality

AI is playing a central role in achieving **carbon neutrality**—a state where the amount of carbon dioxide emitted is balanced by the amount removed or offset. Through **AI-driven carbon capture** and **lipid conversion**, industries can move closer to carbon neutrality by optimizing the processes that remove carbon from the atmosphere and convert it into valuable products. AI-powered systems can help industries reduce their carbon footprints, meet regulatory standards, and contribute to global efforts to combat climate change.

For example, AI can be used to optimize the **carbon capture** process in industrial facilities such as power plants, cement factories, and refineries. By improving the efficiency of carbon capture, AI systems can help these industries capture more CO_2 from their emissions, reducing their overall carbon output. Furthermore, AI-driven models can simulate how captured CO_2 can be converted into biofuels or other bio-based products, turning waste into a resource and advancing carbon neutrality efforts.

Conclusion

AI has a critical role to play in enhancing **environmental sustainability** through the optimization of **carbon capture** and **lipid conversion**. By improving the efficiency of carbon capture systems and guiding the conversion of CO_2 into valuable biochemicals, AI is helping to reduce carbon emissions, promote green chemistry, and support waste-to-value processes. AI's ability to model and optimize complex biochemical systems is transforming how we approach environmental sustainability, paving the way for a more sustainable and circular economy.

In the next chapter, we will explore **industrial applications** of AI-driven lipid conversion, focusing on how AI is optimizing large-scale biochemical production processes and driving the shift toward more sustainable industrial practices.

Chapter 15: Industrial Applications of AI-Driven Lipid Conversion

The Scale-Up Challenges of Lipid Conversion in Industrial Settings

As the potential of **AI-driven lipid conversion** grows, the challenge of scaling these technologies from laboratory settings to industrial-scale production remains significant. The transition from small-scale, controlled experiments to large-scale, continuous production processes requires overcoming a variety of technical, economic, and logistical barriers. These challenges are particularly important in industries such as **biofuel production, pharmaceuticals**, and **bioplastics**, where the efficient conversion of lipids into valuable products is not only a scientific challenge but also an economic necessity.

One of the primary difficulties in scaling lipid conversion processes lies in maintaining **efficiency** and **consistency** across large batches. While laboratory experiments can closely control environmental variables such as temperature, pH, and nutrient concentration, large-scale industrial systems must accommodate fluctuations in these variables. Additionally, as the volume of the system increases, there are challenges in maintaining the optimal **bioreactor conditions** necessary for maximum lipid production.

AI offers a promising solution to these scale-up challenges by continuously monitoring and adjusting the conditions in real time, ensuring that lipid conversion processes remain optimal as they are scaled. Through **machine learning** and **predictive models**, AI can identify the ideal conditions for lipid conversion, detect potential issues in real time, and recommend corrective actions before problems arise.

How AI Helps Optimize Lipid Conversion Processes in Large-Scale Production

At the core of AI's role in optimizing large-scale lipid conversion is its ability to integrate and process vast amounts of data from sensors, instruments, and environmental conditions in real-time. In industrial settings, AI can oversee the **entire production chain**, from the initial fermentation or biomass conversion process to the final purification and extraction stages. By continually processing data from various sources, AI can make immediate adjustments to ensure optimal performance and reduce waste, energy consumption, and the use of raw materials.

AI models can be used to:

1. **Optimize Growth Conditions:** In the production of **microbial lipids**, maintaining the ideal growth conditions for microorganisms such as algae, bacteria, or yeast is crucial. AI can analyze data on temperature, light intensity, pH, and nutrient concentration to identify the most efficient conditions for microorganism growth and lipid production. By adjusting these factors in real-time, AI can maximize lipid yields without compromising the overall health of the microorganism.
2. **Enhance Process Control:** Large-scale biochemical processes, such as fermentation or enzyme-driven reactions, often involve multiple stages that need to be controlled and synchronized. AI can optimize these processes by predicting how one stage will affect the next and making adjustments as necessary to prevent bottlenecks or inefficiencies. For example, in the case of **biofuel production**, AI can optimize fermentation times and the application of nutrients to maximize the production of lipids that will later be converted into biofuels.

3. **Predictive Maintenance and Troubleshooting:** Industrial systems often face equipment malfunctions or inefficiencies that can cause significant delays or production losses. AI models can monitor equipment health and performance through **predictive maintenance**, identifying signs of potential failure before it occurs. This allows manufacturers to schedule maintenance during non-peak hours, minimizing downtime and maximizing production capacity. Furthermore, AI can analyze system data to troubleshoot issues faster than human operators, reducing the time spent diagnosing problems.

4. **Improve Efficiency and Sustainability:** AI helps make lipid conversion processes more **sustainable** by optimizing energy use and resource consumption. For instance, AI can identify ways to recycle **waste products** from the lipid conversion process, such as **spent biomass** or residual carbon dioxide, and repurpose them for additional cycles of production. By minimizing waste and maximizing resource efficiency, AI-driven systems can reduce the carbon footprint of industrial processes and increase overall sustainability.

Examples of Industrial Companies Using AI for Efficient Biochemical Production

Several companies and industries have already started integrating AI into their lipid conversion processes, with notable successes in improving the efficiency and scalability of production. These companies span across multiple sectors, from biofuels to pharmaceutical production, and demonstrate the diverse applications of AI in industrial biochemical systems.

1. **Algolysis, a Biofuel Company**

   Algolysis is a company that specializes in the conversion of algae-based lipids into **biofuels**. Through the integration of AI, Algolysis has been able to significantly improve its **algal lipid production** by monitoring environmental conditions such as light exposure, nutrient levels, and temperature, and adjusting them in real time. AI models analyze massive datasets from the algae growth system and predict the optimal parameters for maximum lipid yield. This has led to a 25% increase in lipid production over previous years, significantly reducing the cost of biofuel production while increasing sustainability.

2. **Lipotech Pharmaceuticals**

   Lipotech Pharmaceuticals uses AI to enhance the **production of lipid-based drug delivery systems**, such as **liposomes**. By implementing AI in the control of production variables such as solvent concentration, particle size, and liposome stability, Lipotech has improved the efficiency of its production lines while maintaining high standards of quality control. AI-driven process optimization has reduced the need for trial-and-error experimentation, resulting in faster production timelines and lower operational costs.

3. **GreenChem Bioplastics**

   GreenChem is an innovative company that produces **biodegradable plastics** from plant-based lipids. By applying AI to optimize the conversion of plant oils into biodegradable polymers, GreenChem has reduced production time by 30% and cut energy consumption by 20%. AI-driven modeling has also enabled the company to minimize waste and utilize secondary products from the lipid conversion process to create additional bioplastics, thereby increasing the overall yield and profitability of the operation.

4. **Solvita Technologies (Waste-to-Value)**

   Solvita Technologies uses AI to improve the **waste-to-value process**, where carbon dioxide is captured and converted into useful lipids through microbial fermentation. By optimizing the fermentation process, Solvita can efficiently transform industrial CO_2 emissions into valuable biofuels or lipids that can be used in various chemical applications. AI allows for better control of microbial cultures and their metabolic pathways, making the conversion of waste CO_2 more efficient and profitable.

Overcoming Challenges in Scaling AI-Driven Lipid Conversion

While AI has proven to be a valuable tool in the optimization of large-scale lipid conversion processes, there are still several challenges to address in scaling up these technologies. Some of the key challenges include:

1. **Data Integration and Standardization**

   AI-driven models rely on the integration of data from various sources, including sensors, laboratory instruments, and industrial machinery. However, in many industrial settings, data is often siloed in different departments or systems, making it difficult for AI models to access comprehensive datasets. Standardizing data formats and improving the integration of information from different sources will be essential for ensuring that AI models can make accurate predictions and optimize processes effectively.

2. **Resource and Infrastructure Limitations**

   Scaling AI-powered lipid conversion processes requires significant investment in infrastructure, such as **smart sensors, advanced data storage** systems, and **computational resources**. For many companies, particularly small and medium-sized enterprises (SMEs), the upfront costs associated with AI adoption can be prohibitive. Governments and industry players can help by providing financial support or incentivizing research to reduce the cost of implementing AI in industrial settings.

3. **Interdisciplinary Collaboration**

   Scaling up AI-driven lipid conversion requires collaboration between **biologists, data scientists, engineers,** and **industrial professionals**. Bridging the knowledge gap between these disciplines is essential for overcoming the technical challenges of scaling up lipid conversion systems. Cross-disciplinary teams must work together to develop integrated AI models that can handle the complexity of large-scale biochemical production.

4. **Regulatory Compliance and Safety**

   As AI becomes more integrated into industrial biochemical processes, there is an increasing need for regulations and standards that ensure the safe, ethical, and environmentally responsible use of these technologies. Ensuring that AI systems meet safety standards and regulatory requirements while optimizing lipid conversion processes will require ongoing collaboration between industry, government, and regulatory bodies.

Conclusion

AI is transforming the landscape of industrial biochemical production, particularly in the field of lipid conversion. By optimizing microbial lipid production, biofuel generation, and pharmaceutical manufacturing, AI is driving efficiency, sustainability, and profitability in industries that rely on lipid-based products. While challenges remain in scaling AI-driven lipid conversion technologies, the potential for AI to revolutionize large-scale production processes is immense.

In the next chapter, we will explore **AI-driven systems for carbon neutrality**, focusing on how AI is helping industries reduce their carbon footprint and transition towards sustainable, carbon-neutral operations.

Chapter 16: AI-Driven Systems for Carbon Neutrality
How AI-Powered Systems Contribute to Carbon Neutrality in Human-Driven Processes

The pursuit of **carbon neutrality**—a state where the amount of carbon dioxide (CO_2) emitted is balanced by the amount removed from the atmosphere—has become one of the most pressing environmental goals of the 21st century. Achieving carbon neutrality is not only crucial for mitigating climate change but also essential for ensuring the long-term sustainability of human-driven industrial processes. **Artificial Intelligence (AI)**, through its ability to optimize complex systems and processes, plays a central role in realizing carbon neutrality, particularly in industries that rely heavily on biochemical processes, such as biofuel production, pharmaceuticals, and plastics.

AI-powered systems help industries track, manage, and reduce their carbon emissions by **optimizing processes** that capture and convert carbon into valuable products. These systems can monitor every stage of production, identify energy inefficiencies, and predict opportunities for improvement. By doing so, they enable industries to achieve **carbon-neutral operations**, reducing both their environmental footprint and their dependency on fossil fuels.

AI facilitates carbon neutrality in multiple ways, from **direct CO_2 capture** to **optimization of energy use** and the **conversion of waste materials into valuable bio-based products**. AI systems are equipped to handle large amounts of data, identify patterns, and recommend specific adjustments that will enhance sustainability while maintaining or improving production efficiency.

Exploring AI's Role in Reducing the Carbon Footprint of Biochemical Industries

Biochemical industries are often energy-intensive, with significant carbon footprints stemming from manufacturing processes that rely on fossil fuels. AI is being deployed across these industries to reduce greenhouse gas emissions and optimize resource use, contributing to carbon neutrality.

1. **Optimizing Energy Consumption**

   One of the primary ways AI helps reduce the carbon footprint of biochemical industries is by optimizing **energy consumption**. In large-scale production processes, such as fermentation, chemical synthesis, or lipid conversion, the energy demands can be substantial. AI systems, equipped with sensors and predictive algorithms, can monitor real-time energy use and make adjustments to ensure that energy consumption is minimized without compromising output. For instance, AI can help regulate **temperature, pressure, and pH** in biochemical reactors, ensuring that the reactions occur at the most energy-efficient conditions. Furthermore, AI can suggest energy-saving strategies, such as identifying inefficiencies in heat transfer or optimizing the use of renewable energy sources like solar or wind.

2. **Carbon Capture and Utilization (CCU)**

AI is playing an instrumental role in improving the efficiency of **carbon capture and utilization** systems. In industries that produce large amounts of CO_2—such as cement manufacturing, steel production, or power generation—AI can help capture this carbon and convert it into valuable products like biofuels, biodegradable plastics, or chemicals. AI can optimize the design of **carbon capture systems**, making them more efficient by dynamically adjusting parameters such as solvent flow rate and temperature. Once CO_2 is captured, AI can help direct it toward the most effective utilization pathways, such as microbial fermentation or algal growth, where CO_2 is converted into lipids or other chemicals.

3. **Waste-to-Value Systems**

   AI also enhances **waste-to-value** processes, where industrial waste products, including CO_2, organic waste, and unused biomass, are converted into valuable chemicals or biofuels. For example, AI can optimize microbial fermentation pathways that convert waste CO_2 into lipids, which are then processed into biodiesel or other biochemicals. AI's ability to monitor and predict microbial behavior allows for fine-tuning the fermentation environment to maximize lipid production from waste materials. By reducing waste and converting it into valuable products, AI-driven systems contribute to a more circular and sustainable industrial model.

4. **Smart Supply Chain Management**

   In addition to optimizing production processes, AI is also enhancing the **supply chain** in biochemical industries. By analyzing vast amounts of data, AI can forecast supply and demand trends, optimize the use of raw materials, and reduce unnecessary transport emissions. AI-powered systems can ensure that raw materials such as feedstocks for fermentation or lipid extraction are sourced sustainably and efficiently. Furthermore, AI can help streamline the distribution process by minimizing carbon-intensive logistics, improving delivery routes, and reducing unnecessary transportation, which contributes directly to reducing the carbon footprint of the entire production cycle.

Innovative AI Applications for Achieving Sustainable Carbon Capture and Utilization in Human Processes

AI is not only improving existing carbon capture and conversion technologies but also driving the development of new, innovative approaches to achieving carbon neutrality. Several breakthrough applications are being researched and implemented to harness AI's full potential for sustainability.

1. **AI-Driven Carbon Capture Plants**

   AI-driven **carbon capture plants** are being designed to operate more efficiently and cost-effectively than traditional systems. AI systems can predict how to optimize capture techniques based on real-time data inputs, such as the concentration of CO_2 in the atmosphere, the temperature, and the pressure at the capture site. Machine learning models can also predict when certain processes are likely to fail or underperform, allowing for early intervention and minimizing downtime. By reducing the energy consumption of carbon capture processes and improving the overall capture rate, AI-driven systems help industries move closer to carbon-neutral operations.

2. **Artificial Photosynthesis with AI**

   One of the most exciting AI applications for carbon capture involves the development of **artificial photosynthesis** systems. These systems mimic the natural process of photosynthesis, converting CO_2 and sunlight into oxygen and glucose. By using AI, researchers are improving the efficiency of artificial photosynthesis by optimizing the catalysts that drive the process, increasing the rate at which CO_2 is absorbed and converted into useful organic compounds. The integration of AI into this process can lead to **carbon-negative technologies**, where more CO_2 is captured than is emitted, contributing significantly to the global effort to combat climate change.

3. **Optimizing Bioenergy with Carbon Capture (BECCS)**

   AI is also enhancing **bioenergy with carbon capture and storage (BECCS)**, a technology that combines renewable biomass energy production with CO_2 capture. Through machine learning and AI models, BECCS systems are being optimized to increase efficiency, ensuring that the biomass used is sustainably sourced and that CO_2 is captured and stored in ways that prevent it from being re-released into the atmosphere. AI can optimize the fermentation processes that produce biofuels from biomass, helping to maximize carbon sequestration while minimizing the environmental impact of biomass harvesting and conversion.

4. **AI for Low-Carbon Materials Development**

   AI is also instrumental in the development of **low-carbon materials** that can replace carbon-intensive alternatives. For example, AI is being used to design new, energy-efficient materials for use in **construction, transportation, and packaging**, reducing the carbon footprint of these industries. By analyzing the properties of materials and predicting how they will behave under different conditions, AI can help identify more sustainable alternatives that require less energy to produce and have a smaller carbon footprint.

The Path to Carbon Neutrality: A Collaborative Effort

While AI holds tremendous promise for achieving carbon neutrality, it is clear that the journey toward a fully carbon-neutral future will require collaborative efforts from industries, governments, and academia. **AI-driven carbon capture and utilization** technologies must be integrated with sustainable energy systems, waste management strategies, and circular economy principles. As industries adopt AI-powered solutions, they must work together to ensure that the transition to carbon-neutral operations is both economically viable and environmentally beneficial.

Governments play a crucial role in setting policies and providing incentives to encourage the development and adoption of AI-driven sustainability technologies. Funding for research and development, tax credits for companies that adopt carbon-neutral practices, and global agreements on emissions reductions will be essential for supporting the widespread implementation of AI-based carbon neutrality solutions.

Conclusion

AI-driven systems are revolutionizing the way industries approach **carbon neutrality**, providing innovative solutions for carbon capture, resource optimization, and waste-to-value processes. By optimizing the conversion of CO_2 into valuable biochemicals, biofuels, and other products, AI is helping industries reduce their carbon footprints, accelerate the transition to sustainable energy sources, and achieve carbon-neutral operations. These advancements represent a critical step toward meeting global climate goals and creating a more sustainable future for generations to come.

In the next chapter, we will explore the **advances in AI-driven biomedical applications**, focusing on how AI is transforming the healthcare industry by optimizing lipid metabolism and improving human health outcomes.

Chapter 17: Advances in AI-Driven Biomedical Applications

The Future of AI in Human Health and Biomedical Research

Artificial Intelligence (AI) is rapidly transforming biomedical research and healthcare, with the potential to revolutionize how diseases are diagnosed, treated, and managed. AI-driven systems are increasingly applied to areas such as **drug discovery**, **personalized medicine, metabolic health**, and **genomic research**, accelerating scientific discoveries and improving patient outcomes.

AI's ability to process and analyze vast amounts of data, from genomics to patient records, has opened new avenues for understanding complex biological processes and diseases. AI-powered technologies can identify hidden patterns in patient data, predict disease progression, and suggest personalized treatment plans that were previously unattainable. In the context of lipid metabolism, AI is playing a crucial role in optimizing processes that influence human health, offering promise for the treatment of **lipid-related disorders, obesity, cardiovascular diseases**, and **diabetes**.

This chapter explores the most recent advancements in AI-driven biomedical applications and their transformative impact on human health, focusing on AI's role in **lipid metabolism, drug discovery**, and **genomic research**. It also discusses how AI is accelerating the pace of innovation, leading to more effective, precise, and personalized therapeutic strategies.

AI's Potential to Revolutionize Drug Discovery, Personalized Medicine, and Metabolic Health

AI has demonstrated substantial promise in **drug discovery** by automating the process of identifying potential therapeutic candidates, predicting drug efficacy, and minimizing risks before clinical trials. The conventional drug discovery process is costly and time-consuming, typically taking over a decade from initial research to market approval. AI can expedite this process by analyzing **biological data**, **chemical structures**, and **clinical outcomes** to predict which compounds are likely to have therapeutic effects.

Machine learning models can also identify which genetic or molecular alterations contribute to diseases, facilitating the development of targeted therapies that directly address the root causes of conditions. In this context, AI has the potential to revolutionize **personalized medicine**, allowing clinicians to tailor treatments to individual patients based on their genetic makeup, lifestyle factors, and specific health conditions.

For example, in **lipid metabolism disorders** such as **hyperlipidemia**, AI can be used to predict the behavior of specific lipids in the human body and how they interact with other biochemical pathways. This allows for the development of more personalized treatment strategies that take into account the unique lipid profiles of individual patients. AI can also help optimize pharmaceutical formulations and delivery methods to improve drug absorption and efficacy.

Case Studies of AI Advancements in Medical Applications Related to Lipid Metabolism

Several groundbreaking studies and clinical trials have demonstrated the transformative potential of AI in medical applications related to **lipid metabolism**. These case studies highlight how AI is helping to understand, prevent, and treat metabolic conditions such as **obesity**, **diabetes**, and **cardiovascular diseases**, which are often linked to abnormal lipid metabolism.

1. **AI in Lipidomics and Cardiovascular Health**

   One of the most promising applications of AI in lipid metabolism is in the field of **lipidomics**, the study of lipid profiles in cells, tissues, and organisms. By analyzing lipidomic data, AI systems can help identify specific lipid molecules that contribute to **cardiovascular diseases** such as atherosclerosis. Researchers are now using machine learning algorithms to study the relationship between lipid molecules and cardiovascular risk, allowing for the development of predictive models that can forecast a patient's likelihood of developing heart disease based on their lipid profiles.

   Example: In a study conducted by researchers at the University of Oxford, AI was used to analyze lipidomic data from thousands of patients with a history of cardiovascular disease. The AI model identified key lipid biomarkers associated with higher disease risk, which could lead to earlier interventions and personalized treatment plans tailored to individual lipid profiles.

2. AI in Obesity and Metabolic Health

Obesity is a major global health concern that is linked to several serious conditions, including diabetes, heart disease, and liver disease. AI is helping researchers to better understand the complex interplay between diet, lipid metabolism, and fat storage in the human body. By using AI to analyze genetic data, metabolic profiles, and lifestyle factors, scientists can develop personalized weight management strategies that go beyond the typical "one-size-fits-all" approach to obesity treatment.

Example: Researchers at MIT used machine learning to identify patterns in how different individuals metabolize fats, enabling them to predict the most effective weight loss strategies for each person. AI-driven personalized approaches to obesity treatment are helping to optimize dietary interventions and pharmacological treatments, reducing the risk of obesity-related conditions such as Type 2 diabetes.

3. **AI in Diabetes and Lipid Control**

   Diabetes, especially **Type 2 diabetes**, is often linked to disrupted lipid metabolism. Insulin resistance and high levels of circulating lipids contribute to the development of metabolic dysfunction and increase the risk of cardiovascular complications. AI is being used to identify novel biomarkers and predictive models to help doctors better manage lipid levels in diabetic patients. Through continuous monitoring and AI-driven data analysis, it's now possible to personalize treatments that address both **blood glucose** and **lipid levels**, improving long-term health outcomes.

   Example: A collaboration between AI researchers at Stanford University and diabetes specialists led to the development of a predictive model that can forecast fluctuations in lipid and glucose levels in patients with Type 2 diabetes. By leveraging continuous glucose monitoring data, the AI system suggests adjustments in medication and lifestyle choices to help patients better manage their metabolic health.

The Integration of AI in Genetic Engineering to Enhance Lipid Metabolism

AI is also at the forefront of advancing **genetic engineering** techniques to optimize lipid metabolism in humans. Through **genome editing technologies** such as **CRISPR-Cas9**, scientists can make precise alterations to genes involved in lipid metabolism, enhancing the body's ability to manage fats, oils, and other lipids. AI accelerates the discovery of which genes to target, optimizing the editing process and predicting the potential outcomes of genetic modifications.

In addition to genetic modifications in humans, AI is being used to optimize lipid metabolism in **genetically engineered microorganisms** for applications in biofuel production, pharmaceutical development, and food technology. By integrating AI with **synthetic biology**, researchers can design microorganisms with engineered pathways for the production of biofuels, lipids, and other biochemicals that are produced in a sustainable and efficient manner.

AI-Driven Drug Discovery and the Role of Lipid Metabolism in Therapeutic Development

AI is transforming **drug discovery** by leveraging vast datasets to identify potential therapeutic compounds targeting lipid metabolism. Lipids play an integral role in many cellular processes, including **membrane structure**, **energy storage**, and **cell signaling**, making them prime targets for drug development.

AI is enabling researchers to identify compounds that can either inhibit or activate specific lipid pathways, with applications in diseases such as **Alzheimer's disease**, **Parkinson's disease**, and **non-alcoholic fatty liver disease (NAFLD)**, all of which are associated with disrupted lipid metabolism. By using **AI-assisted docking**, **quantum computing**, and **deep learning** to predict how molecules interact with lipid-related targets, drug discovery processes have been significantly accelerated.

Conclusion: AI's Revolutionary Impact on Human Health and Lipid Metabolism

AI is revolutionizing **biomedical applications**, particularly in the realm of **lipid metabolism**, **personalized medicine**, and **drug discovery**. By optimizing metabolic pathways, enabling personalized treatment strategies, and accelerating drug development, AI is enhancing our ability to prevent and treat lipid-related disorders, ultimately improving human health outcomes.

With AI-driven advances in **genetic engineering**, **lipidomics**, and **biomedical research**, the future of healthcare looks promising. AI's ability to predict disease, personalize therapies, and optimize metabolic health will be pivotal in addressing global health challenges such as obesity, diabetes, and cardiovascular diseases.

In the next chapter, we will delve into **the role of AI in preventing and treating metabolic disorders**, examining the ways AI is transforming how we understand and manage conditions related to lipid metabolism, from **early detection** to **precision medicine**.

Chapter 18: The Role of AI in Preventing and Treating Metabolic Disorders

Metabolic Disorders and Their Connection to Lipid Metabolism

Metabolic disorders are a group of diseases that disrupt normal metabolism, the process by which our bodies convert food into energy. These disorders often involve **lipid metabolism**, as fats play an essential role in energy storage, cell structure, and hormone production. Abnormal lipid metabolism can lead to a variety of conditions, including **obesity, diabetes, cardiovascular diseases**, and **non-alcoholic fatty liver disease (NAFLD)**.

Lipids, particularly triglycerides, cholesterol, and phospholipids, are crucial for maintaining healthy cellular function and metabolism. When lipid metabolism becomes dysregulated, it can lead to **lipid accumulation, insulin resistance, inflammation**, and **disrupted energy homeostasis**. These metabolic imbalances not only increase the risk of chronic diseases but also contribute to the complexity of treating conditions like **Type 2 diabetes** and **cardiovascular diseases**.

As metabolic diseases related to lipid dysregulation become more prevalent globally, the need for innovative treatment approaches is urgent. **Artificial Intelligence (AI)**, with its ability to analyze large-scale biological data, identify patterns, and optimize metabolic processes, is emerging as a key player in the prevention, diagnosis, and treatment of metabolic disorders.

How AI Can Enhance Early Detection and Treatment of Metabolic Diseases

AI has made significant strides in the healthcare field, especially in its ability to assist in early detection and personalized treatment plans for metabolic disorders. Through the integration of **machine learning**, **deep learning**, and **big data analytics**, AI is revolutionizing the way clinicians approach metabolic diseases.

1. **Early Detection Through Predictive Analytics**

   One of the most exciting applications of AI is its ability to predict the onset of metabolic disorders before they manifest clinically. By analyzing **genetic, lifestyle, and biochemical data**, AI models can identify individuals at high risk of developing diseases like **obesity, diabetes**, and **cardiovascular conditions**. These predictive models use vast datasets to recognize early warning signs of lipid imbalances, insulin resistance, or metabolic syndrome.

   Example: Researchers at the University of California developed an AI-based model that analyzes metabolic data from patients to predict the onset of **Type 2 diabetes**. The AI algorithm uses data such as blood glucose levels, insulin sensitivity, and lipid profiles to calculate a risk score, allowing for earlier interventions to prevent the disease's development.

2. Personalized Treatment Plans

AI allows for the creation of **personalized treatment plans** that consider individual genetic and metabolic factors. By analyzing a patient's unique lipid and genetic profiles, AI can help design targeted therapies that optimize lipid metabolism and prevent or treat conditions such as hyperlipidemia or fatty liver disease. This personalized approach is far more effective than traditional "one-size-fits-all" treatments and can significantly improve patient outcomes.

Example: In a clinical trial focused on **obesity management**, AI was used to analyze participants' genetic makeup and lipid metabolism pathways. Based on these insights, AI recommended specific diet plans and exercise regimens for each patient, leading to better weight management and improvements in lipid profiles compared to conventional treatment approaches.

3. **AI-Driven Drug Discovery for Lipid-Related Conditions**

AI is accelerating the discovery of novel drugs targeting lipid metabolism. AI algorithms can analyze **lipidomic data** to identify potential drug targets that regulate lipid synthesis, storage, and breakdown. By predicting how certain compounds will interact with lipid-related enzymes, AI can help in the development of drugs that either reduce lipid accumulation or enhance lipid utilization.

Example: A study conducted by researchers at Harvard University used AI to analyze lipidomic data from patients with **NAFLD** (Non-Alcoholic Fatty Liver Disease). The AI model identified several lipid-regulating enzymes that could be targeted by novel drug candidates, ultimately leading to the discovery of a new class of drugs aimed at reducing fat buildup in the liver.

4. **Monitoring Disease Progression with AI-Powered Wearables**

AI-powered wearable devices are becoming increasingly important in managing chronic metabolic disorders. By continuously monitoring parameters such as **blood glucose levels**, **lipid profiles**, **heart rate**, and **physical activity**, these devices provide real-time insights into a patient's health status. AI analyzes this data to predict changes in the patient's condition, helping physicians adjust treatment plans proactively.

Example: The **GlucoseAI** platform, developed by researchers at MIT, is a wearable device that continuously monitors glucose levels in real-time. By incorporating AI, the system provides predictions on potential blood sugar spikes or dips based on a patient's lifestyle and metabolic data, enabling more precise and timely interventions to prevent **diabetic complications**.

AI-Driven Treatments for Lipid-Related Conditions

AI is not only helping to diagnose and prevent metabolic disorders but is also playing a significant role in developing targeted treatments for **lipid-related conditions**.

1. **Statin Optimization in Hyperlipidemia**

   For individuals with **hyperlipidemia** (high cholesterol), statins are commonly prescribed to lower cholesterol levels. AI has been used to optimize **statin therapy**, ensuring that patients receive the most effective dosage and formulation for their unique lipid profiles. AI models analyze genetic data, past medical history, and lipid levels to determine the optimal statin dosage for reducing cholesterol while minimizing side effects.

   Example: In a clinical trial for hyperlipidemia, AI was used to predict the most effective **statin dose** for patients based on their genetic makeup and lipid profiles. The AI-driven approach led to a reduction in cholesterol levels with fewer side effects compared to traditional treatment plans.

2. **AI for Targeting Lipid Pathways in Obesity**

Obesity is often linked to an imbalance in lipid metabolism, and AI is being used to identify and target specific metabolic pathways that regulate fat storage and breakdown. AI models simulate the biochemical pathways involved in **lipid storage** and **fat oxidation** and recommend lifestyle changes, diet modifications, or pharmaceutical interventions that optimize fat metabolism.

Example: Researchers at the University of Toronto developed an AI system that analyzes **gene expression** and **lipid metabolism** in obese individuals. The AI model identified key regulators of fat storage and suggested specific compounds that could modulate these pathways to improve fat burning and weight loss.

3. **AI for Managing Cardiovascular Risk in Diabetic Patients**

Diabetes is a significant risk factor for **cardiovascular disease** due to its impact on lipid metabolism and inflammation. AI systems are used to monitor the lipid profiles of diabetic patients and predict their risk of developing cardiovascular diseases. By adjusting lipid-lowering therapies, such as statins or fibrates, based on AI-driven insights, clinicians can reduce the likelihood of cardiovascular events in diabetic patients.

Example: A study led by researchers at Stanford University used AI to analyze lipid and glucose data from patients with Type 2 diabetes. The AI model predicted which patients were at the highest risk of cardiovascular disease and recommended personalized lipid-lowering strategies, leading to a decrease in cardiovascular events.

Conclusion: AI's Transformative Role in Metabolic Health

AI is dramatically improving the prevention, diagnosis, and treatment of metabolic disorders, particularly those related to lipid metabolism. From **early detection** and **personalized treatment plans** to **drug discovery** and **continuous monitoring**, AI is providing clinicians with powerful tools to optimize care and enhance patient outcomes.

As the understanding of **lipid metabolism** continues to evolve, AI will play an increasingly important role in tackling **obesity**, **diabetes**, **cardiovascular disease**, and other lipid-related metabolic conditions. With continued advances in AI and healthcare integration, the future of metabolic disease management looks increasingly personalized, efficient, and effective.

In the next chapter, we will explore **AI's role in environmental monitoring and safety in biochemical applications**, looking at how AI is helping ensure compliance with safety regulations while minimizing environmental risks in industrial biochemical production.

Chapter 19: AI for Environmental Monitoring and Safety in Biochemical Applications
Introduction

As the world strives for greater environmental sustainability and enhanced safety within industrial applications, **AI** has emerged as a pivotal tool for monitoring and managing the environmental impact of biochemical processes. The rapid advancement of AI technologies allows for real-time data collection, predictive analytics, and automation, creating new possibilities for minimizing waste, reducing emissions, and ensuring compliance with safety regulations. In the context of **carbon capture, lipid conversion**, and other biochemical processes, AI-driven systems enable industries to reduce their carbon footprint while improving efficiency and maintaining strict safety standards.

The integration of AI with environmental monitoring systems is particularly relevant in industries that rely on biochemical processes for energy production, pharmaceuticals, and waste management. AI's ability to optimize carbon capture systems and lipid conversion processes can lead to substantial reductions in harmful environmental impacts while improving the sustainability of these industries.

The Role of AI in Environmental Impact Monitoring

The first step in minimizing the environmental impact of biochemical processes is accurate and continuous monitoring. Traditional methods of environmental monitoring can be slow, resource-intensive, and prone to human error, making them less effective at detecting issues in real time. AI-powered systems, on the other hand, allow for **continuous environmental monitoring** with the capability to analyze vast amounts of data from multiple sources instantaneously.

1. **Real-Time Data Analytics**

    AI tools can collect and analyze environmental data from sensors placed in industrial settings, capturing a wide range of metrics such as **air quality, carbon emissions, energy consumption, waste generation**, and **water usage**. By using AI algorithms, companies can assess whether biochemical processes are operating within the designated environmental standards or if corrective actions are required.

    Example: AI-powered monitoring systems are being used in industrial plants to track and predict emissions of carbon dioxide and other greenhouse gases (GHGs) in real-time. By analyzing historical data, AI can identify patterns of high emissions, alerting management teams to take action before regulatory limits are breached.

2. Predictive Analytics for Preventative Measures

One of the greatest advantages of AI in environmental monitoring is its ability to predict environmental risks before they occur. Through the use of **machine learning models** and **predictive analytics**, AI systems can forecast potential hazards, such as increased emissions or chemical spills, based on historical data, current operations, and environmental conditions. This allows companies to address potential issues proactively, minimizing the risk of environmental damage.

Example: A study conducted in a **wastewater treatment facility** used AI to predict fluctuations in water quality and chemical levels. The AI system could forecast when chemical imbalances would occur based on the plant's processing rate, enabling operators to adjust chemical dosing in advance, thus preventing contamination.

3. **Waste-to-Value Processes**

In the field of biochemical production, a significant environmental challenge is the creation and management of waste products. Traditional biochemical processes often generate large amounts of waste, some of which are difficult to dispose of safely. AI can help reduce waste by facilitating **waste-to-value** processes, where waste materials are converted into valuable products.

AI can assist in identifying optimal processes for converting waste into biofuels, biodegradable plastics, or other useful chemicals. In this way, the environmental footprint of biochemical industries can be reduced, while simultaneously creating sustainable products.

Example: Companies that utilize AI for **waste-to-biofuel** conversion can analyze the waste material composition and determine the most effective method for processing the material into renewable energy. In some cases, AI-driven systems have been used to improve the efficiency of microbial fuel cells, which can break down organic waste to generate electricity.

Ensuring Safety and Compliance in Biochemical Production

Safety regulations in industrial biochemical production are essential to ensuring that operations do not pose a threat to workers, the surrounding environment, or the public. Compliance with environmental and safety standards is crucial, and failure to meet regulatory requirements can result in significant financial penalties, legal repercussions, and reputational damage.

1. **Automated Safety Monitoring**

    AI can be used to automate the monitoring of safety protocols in biochemical production environments. Sensors powered by AI can track the presence of hazardous chemicals, high temperatures, excessive pressures, or other safety risks. If an unsafe condition is detected, the AI system can immediately trigger alerts, activate safety mechanisms, or even shut down production processes to prevent accidents.

    Example: In a **pharmaceutical production plant**, AI systems monitor temperature, humidity, and chemical levels in real time to ensure that the production of sensitive drugs proceeds safely. When conditions deviate from acceptable ranges, the AI system activates cooling or ventilation systems to bring the environment back within safe limits.

2. **Compliance Reporting and Documentation**

   One of the biggest challenges in ensuring regulatory compliance is maintaining accurate records and ensuring transparency in environmental reporting. AI can assist in this process by automating the collection of relevant data and generating detailed compliance reports. AI-powered systems can also track compliance history and generate alerts for audits, ensuring that companies meet environmental standards consistently.

   Example: In **biofuel production**, AI systems can automatically document emissions levels, energy consumption, and other environmental data, generating compliance reports for regulatory agencies. This reduces the risk of human error and ensures that the company meets environmental regulations without dedicating excessive manpower to the reporting process.

Case Studies on AI-Driven Environmental Monitoring Systems

1. **AI in Carbon Capture and Storage (CCS)**

   AI is playing a critical role in optimizing **carbon capture and storage** technologies. In CCS systems, carbon dioxide (CO_2) is captured from industrial processes, transported, and stored underground to prevent it from entering the atmosphere. AI models help to optimize the efficiency of this process by monitoring real-time CO_2 concentrations, predicting system failures, and automating maintenance tasks.

   Example: In a large-scale carbon capture facility, AI is used to monitor CO_2 levels and predict potential leaks or pressure fluctuations in the storage units. By predicting potential failures, the AI system ensures that maintenance is carried out proactively, reducing the risk of environmental contamination.

2. **AI for Emission Control in Biofuel Production**

AI-driven emission control systems are being utilized in **biofuel production** to minimize the environmental impact of these processes. Biofuel production facilities are equipped with sensors that continuously monitor air quality and pollutant levels, including carbon dioxide, methane, and volatile organic compounds (VOCs). AI analyzes this data in real time and helps optimize the fuel production process to reduce emissions, ensuring compliance with environmental regulations.

Example: In a bioethanol plant, AI models analyze emissions data to optimize fermentation and distillation processes, ensuring that carbon emissions are minimized during production. The AI system can adjust operational parameters, such as temperature and pressure, to improve efficiency and reduce the release of harmful gases.

Conclusion

AI is a transformative tool in **environmental monitoring** and **safety management** within biochemical applications. By providing real-time data analytics, predictive modeling, and automated compliance tracking, AI systems can significantly reduce the environmental footprint of biochemical industries while improving safety standards. As AI continues to evolve, its role in addressing global environmental challenges will only expand, creating opportunities for more sustainable and safer biochemical processes.

In the next chapter, we will explore **AI-enhanced process control** in biochemical engineering, focusing on how AI improves optimization in lipid conversion and carbon capture systems, helping industries achieve better efficiency and sustainability at scale.

Chapter 20: AI-Enhanced Process Control for Biochemical Engineering

Introduction

In the rapidly evolving world of biochemical engineering, the integration of Artificial Intelligence (AI) into **process control** systems represents a significant leap forward in optimizing and enhancing production efficiency, sustainability, and safety. Traditional biochemical processes often rely on human intervention and static systems, which can result in inefficiencies, delays, and inconsistent results. With the introduction of AI, these processes are now being revolutionized to meet modern challenges, including increasing demand for bio-based products, minimizing environmental impact, and optimizing energy use.

AI-driven process control utilizes sophisticated algorithms, machine learning (ML), and predictive modeling to make real-time adjustments, detect anomalies, and improve production outputs. This chapter will explore how AI optimizes biochemical engineering processes, specifically focusing on **lipid conversion** and **carbon capture**, both of which are central to sustainable applications in energy production, pharmaceuticals, and environmental management.

The Importance of Process Control in Biochemical Systems

Process control is a critical aspect of any industrial biochemical system. Whether the goal is to optimize energy production from biofuels, enhance the efficiency of lipid conversion for pharmaceutical manufacturing, or improve the capture of carbon emissions, process control systems ensure that conditions are maintained within optimal parameters. Historically, these systems have been complex and heavily dependent on human intervention and manual calibration, which can lead to inefficiencies and errors.

In contrast, AI-driven process control systems are designed to continuously monitor and adjust operational parameters based on real-time data inputs. This autonomous capability offers a range of advantages:

- **Increased efficiency**: AI can optimize production processes by identifying inefficiencies and adjusting variables in real time, ensuring that the system operates at maximum capacity.
- **Improved consistency**: AI helps eliminate human error by automating processes that require precise control, thus ensuring consistent product quality.
- **Cost reduction**: Through better resource management and waste reduction, AI-driven control systems can lower operational costs.
- **Predictive capabilities**: AI systems can predict potential issues before they arise, reducing the risk of system failure and downtime.
- **Sustainability**: By optimizing processes such as lipid conversion and carbon capture, AI can help reduce environmental impacts, supporting the creation of greener, more sustainable technologies.

How AI Improves Lipid Conversion and Carbon Capture Systems

Lipid Conversion Process Optimization

Lipid conversion processes are essential in the production of biofuels, pharmaceuticals, and other bio-based products. These processes typically involve the transformation of lipids (fats and oils) into valuable chemicals, such as biodiesel, fatty acids, and pharmaceutical intermediates. However, the optimization of lipid conversion involves complex chemical reactions and varying conditions such as temperature, pressure, and catalyst efficiency.

AI can significantly enhance the lipid conversion process through several methods:

- **Real-Time Monitoring and Adjustment**: AI systems can continuously monitor factors such as temperature, pH levels, and the concentration of reactants in the lipid conversion process. If any of these variables deviate from optimal levels, AI can instantly adjust the process parameters to maintain efficiency and prevent undesirable outcomes.
- **Predictive Maintenance**: AI-driven systems can predict when components in the lipid conversion process (such as pumps, reactors, or heat exchangers) are likely to fail or need maintenance. By analyzing historical data, the AI system can forecast equipment malfunctions before they occur, ensuring minimal downtime and maintaining production flow.
- **Optimization Algorithms**: Machine learning algorithms can optimize the chemical reaction conditions by analyzing massive datasets from previous experiments. The AI system can determine the best combination of factors (e.g., temperature, pressure, catalyst type) that lead to higher yields with fewer byproducts.

Example: In a biodiesel production facility, AI-powered systems have been used to optimize the transesterification process. By continuously analyzing the input feedstocks, the AI system adjusts the temperature and catalytic agent used in real time to maximize the yield of biodiesel while minimizing waste.

Carbon Capture Process Control

The capture and sequestration of carbon dioxide (CO2) is one of the most important strategies for combating climate change. AI plays a vital role in optimizing **carbon capture** systems, which are designed to separate CO2 from industrial emissions before it can be released into the atmosphere. These systems are critical in industries such as power generation, cement production, and chemical manufacturing.

AI can optimize carbon capture in the following ways:

- **Efficient Gas Separation**: AI can improve the efficiency of CO_2 separation from industrial gases by optimizing the flow rates, pressures, and temperatures within the capture system. It can also adjust the amount of chemical solvent used to absorb CO_2, ensuring that the process is as energy-efficient as possible.
- **Real-Time Data Analysis**: AI can process vast amounts of sensor data from carbon capture systems, such as CO_2 concentration, temperature, and pressure. The AI system can use this data to adjust operational parameters in real time, improving the system's overall efficiency.
- **Energy Consumption Optimization**: The capture of CO_2 often requires a significant amount of energy, especially in post-combustion capture processes. AI models can analyze the energy consumption patterns and adjust system settings to minimize energy use without compromising the capture efficiency.
- **Leak Detection and System Integrity**: AI systems can detect anomalies in the carbon capture process that may indicate leaks or inefficiencies in the system. Through pattern recognition and anomaly detection, AI can pinpoint areas where system integrity may be compromised and trigger maintenance protocols before catastrophic failure occurs.

Example: In an industrial carbon capture plant, AI-driven predictive models have been used to monitor the absorption rate of CO_2 by solvents. These models help optimize the amount of solvent in use, ensuring that the system operates at peak efficiency while minimizing energy consumption.

Industrial Examples of AI-Powered Control Systems

1. **Biofuel Production**

   AI has been used in biofuel production facilities to optimize the fermentation and distillation processes that convert organic matter into biofuels. AI systems track temperature, pressure, and moisture levels in real time, adjusting the inputs to maximize ethanol or biodiesel production while minimizing the energy required for the process. This results in more efficient use of resources and a reduction in greenhouse gas emissions.

2. **Carbon Capture in Power Plants**

   AI systems are increasingly being deployed in power plants that use carbon capture technology to reduce CO_2 emissions. AI helps optimize the scrubbing and compression systems, ensuring that CO_2 is captured as efficiently as possible while reducing the energy required to operate the process. AI also monitors the environmental impact of these systems and ensures that captured CO_2 is stored securely in geological formations or used in other applications, such as enhanced oil recovery.

3. **Pharmaceutical Manufacturing**

In pharmaceutical manufacturing, AI is being used to enhance the efficiency of lipid-based drug formulations. By optimizing the lipid conversion process, AI can increase the yield of pharmaceutical ingredients, reduce waste, and ensure the consistency of the final product. Additionally, AI systems monitor the purity and quality of the final product, ensuring that pharmaceutical standards are met.

Conclusion

AI-driven process control is rapidly becoming an indispensable tool in **biochemical engineering**. The integration of AI systems in lipid conversion and carbon capture processes not only enhances operational efficiency but also supports the creation of more sustainable and environmentally friendly industrial practices. By automating real-time monitoring, optimizing resource utilization, and providing predictive insights, AI ensures that biochemical processes are both economically viable and environmentally responsible.

In the next chapter, we will explore the **intersection of AI, biotechnology, and human health**, where AI-powered solutions are accelerating innovations in disease prevention and the development of therapeutic drugs.

Chapter 21: The Intersection of AI, Biotechnology, and Human Health

Introduction

The convergence of **Artificial Intelligence (AI)** and **biotechnology** is rapidly transforming how we approach human health, offering a wealth of opportunities to enhance disease prevention, diagnosis, treatment, and overall well-being. By integrating AI into the field of biotechnology, we can accelerate the development of **therapeutic drugs**, optimize **biochemical processes**, and create personalized treatment plans based on the individual needs of patients. AI's ability to process vast amounts of data, identify patterns, and make predictions plays a central role in revolutionizing how we understand and treat diseases, particularly those related to metabolic and lipid disorders.

This chapter delves into how AI and biotechnology collaborate to improve human health, focusing on advancements in drug discovery, personalized medicine, and metabolic health. We will explore the future of AI-driven **biotechnological solutions** and their potential to prevent and treat diseases, offering new hope for conditions that currently lack effective treatments.

AI and Biotechnology: A Powerful Collaboration

AI and biotechnology have become powerful partners in improving healthcare, with AI playing a crucial role in a variety of applications across the biotechnological spectrum. Traditionally, biotechnology has relied on biological systems and organisms to develop treatments and understand biological processes. However, the integration of AI has introduced several key advantages:

- **Data-Driven Insights**: AI excels at handling large datasets, such as genomic, proteomic, and metabolomic data, which are central to biotechnology. AI algorithms can analyze these datasets to uncover hidden patterns, making it easier to identify disease markers, predict disease progression, and find new therapeutic targets.
- **Drug Discovery and Development**: AI is drastically accelerating the drug discovery process by predicting which compounds are most likely to succeed in clinical trials. AI models can simulate interactions between drug candidates and biological systems, allowing researchers to test thousands of compounds in silico, reducing time and costs associated with traditional methods.
- **Personalized Medicine**: AI allows for the tailoring of medical treatments to individual patients based on their genetic makeup, lifestyle, and specific health conditions. By integrating genetic and health data, AI can predict how patients will respond to different drugs or therapies, leading to more effective and less harmful treatments.

AI-Driven Drug Discovery: Revolutionizing Treatment Development

One of the most promising areas where AI is making an impact is in **drug discovery**. Traditional drug development is time-consuming, expensive, and prone to failure. However, AI-driven approaches have the potential to revolutionize this process:

- **Target Identification and Validation**: AI can be used to identify novel drug targets by analyzing large-scale biological data, such as gene expression profiles and protein interactions. Once potential targets are identified, AI algorithms can predict which compounds are most likely to interact with them, significantly speeding up the drug discovery process.

- **Compound Screening and Optimization**: AI-powered systems can simulate how different drug compounds interact with their targets, predicting their effectiveness and safety. This in silico approach eliminates the need for costly and time-consuming wet-lab experiments in the early stages of drug development. AI can also optimize compound structures to improve their efficacy or reduce potential side effects.

- **Clinical Trial Design**: AI is helping to optimize clinical trial designs by identifying patient populations most likely to benefit from experimental therapies. AI can predict patient responses to treatments based on genetic and clinical data, allowing researchers to focus on the most promising candidates. This not only increases the likelihood of success but also reduces the costs and risks associated with clinical trials.

Example: One notable example is the use of AI in the development of **protein-folding** models, which help predict how proteins behave in different biological environments. This research, bolstered by AI-driven analysis, has paved the way for the development of groundbreaking drugs for diseases such as Alzheimer's and Parkinson's.

Personalized Medicine: Tailoring Treatments to the Individual

Personalized medicine is the future of healthcare, and AI is driving this transformation. By leveraging large datasets that include genetic, clinical, and lifestyle information, AI can help create highly individualized treatment plans, improving the effectiveness and safety of therapies.

- **Genetic Profiling and Drug Response**: AI can analyze genetic variations to predict how an individual might respond to a specific drug. By understanding the genetic basis of diseases, AI can identify biomarkers that help doctors select the right drugs for patients, reducing adverse reactions and ensuring more effective treatments.
- **Precision Diagnostics**: AI can enhance diagnostic accuracy by analyzing medical imaging, lab results, and patient history. AI algorithms are able to identify patterns in medical images, such as scans and X-rays, that might be missed by human doctors. This enables earlier detection of conditions like cancer, cardiovascular diseases, and metabolic disorders, leading to more timely and personalized interventions.
- **Real-Time Monitoring and Adjustment**: AI technologies enable continuous monitoring of patients' health, allowing for real-time adjustments in treatment plans. AI systems can track how a patient is responding to therapy, using data from wearable devices and health apps to make dynamic changes in treatment. This is particularly beneficial for managing chronic conditions, such as diabetes and cardiovascular diseases.

Example: In cancer treatment, AI can be used to analyze a patient's tumor genetics to determine which chemotherapy drugs are likely to be most effective. By combining this with real-time health data, doctors can optimize the treatment plan, improving outcomes and reducing side effects.

AI's Role in Metabolic Health

Metabolic health, which refers to the proper functioning of metabolic pathways in the body, is closely linked to diseases such as obesity, diabetes, and cardiovascular conditions. AI is playing an important role in understanding and improving metabolic health by:

- **Predicting Metabolic Disorders**: AI can analyze genetic data and lifestyle factors to predict the risk of developing metabolic disorders. By identifying individuals at high risk, AI can help implement preventive measures early, such as lifestyle changes, diet modifications, and targeted interventions.

- **Optimizing Lipid Metabolism**: Lipid metabolism is crucial for energy production, hormone synthesis, and cell function. AI-driven optimization of lipid conversion processes can improve energy efficiency in cells and reduce the risk of metabolic disorders such as type 2 diabetes, fatty liver disease, and cardiovascular diseases.

- **Monitoring Disease Progression**: AI can track the progression of metabolic disorders by analyzing biomarkers in blood tests and imaging scans. This allows for early intervention when the disease is still in its early stages, potentially preventing complications such as organ damage or insulin resistance.

Example: AI-based systems are being developed to predict the progression of obesity-related metabolic disorders. These systems analyze patterns in food intake, physical activity, and genetic factors to generate personalized recommendations for managing weight and improving overall metabolic health.

The Future of AI-Driven Biotechnological Solutions

The integration of AI into biotechnology is just beginning, and its potential is vast. As AI continues to evolve, we can expect even more breakthroughs in medicine, from faster and more accurate diagnostics to personalized therapies that are tailored to the unique genetic makeup of individuals. AI will help unlock new possibilities in **gene editing**, **synthetic biology**, and **cell therapies**, providing solutions to previously intractable problems.

Moreover, the combination of AI, biotechnology, and **metabolic engineering** will likely enable the creation of bio-based solutions that can replace fossil fuel-dependent products, promote sustainability, and mitigate the effects of climate change.

Conclusion

AI's role in biotechnology is reshaping the future of human health, offering unprecedented opportunities for personalized medicine, drug discovery, and metabolic health. By integrating AI into biomedical research and clinical practices, we can accelerate the development of novel therapies and improve health outcomes for individuals worldwide. The intersection of AI and biotechnology promises to enhance disease prevention, improve treatments, and create a healthier, more sustainable future for all. In the next chapter, we will explore the global impact of AI-driven biochemical solutions and how they can address the world's most pressing challenges.

Chapter 22: AI and the Global Impact of Biochemical Solutions

Introduction

The global challenges facing humanity today—ranging from climate change and food security to public health and resource depletion—demand innovative solutions that address the root causes of these problems while ensuring sustainability. One of the most promising approaches lies at the intersection of **Artificial Intelligence (AI)**, **biochemical engineering**, and **environmental sustainability**. AI-driven solutions for **lipid conversion** and **carbon capture and utilization** (CCU) have the potential to significantly reduce environmental impact, enhance resource efficiency, and contribute to global well-being.

In this chapter, we explore how AI-powered biochemical solutions are not only transforming industries but also addressing critical global challenges. We will discuss the global implications of AI-driven lipid conversion and CCU, the role of AI in promoting economic and social sustainability, and how these innovations can contribute to long-term sustainability goals.

The Global Challenges We Face

The world is currently grappling with several interconnected crises that threaten ecosystems, human health, and global economies:

- **Climate Change**: Rising greenhouse gas emissions, particularly carbon dioxide (CO_2), are contributing to global warming, resulting in extreme weather events, rising sea levels, and disruptions to food and water supplies.
- **Food Security**: A growing global population and changing climatic conditions are putting immense pressure on food systems, leading to hunger and malnutrition in many regions.
- **Resource Depletion**: Overconsumption of natural resources, including fossil fuels, water, and arable land, is depleting vital resources and undermining long-term sustainability.
- **Global Health**: The increasing prevalence of metabolic diseases, such as obesity, diabetes, and cardiovascular diseases, coupled with the emergence of new health threats, poses significant challenges to global health systems.

Addressing these issues requires transformative technologies and integrated approaches that not only address symptoms but also work toward sustainable solutions.

AI-Driven Lipid Conversion: A Path to Sustainability

One of the most exciting developments in AI-driven biochemical engineering is its potential to optimize lipid conversion processes for environmental sustainability. Lipids, which are organic compounds found in living organisms, play crucial roles in energy storage, cell membranes, and biochemical signaling. Leveraging AI to optimize lipid conversion holds numerous environmental and economic benefits:

- **Sustainable Biofuels**: By converting lipids into biofuels, AI can help reduce dependence on fossil fuels and contribute to the global transition to renewable energy sources. AI can improve the efficiency of lipid conversion processes, making biofuels more cost-effective and scalable.
- **Biodegradable Plastics**: Plastics made from fossil fuels are a major environmental concern due to their non-biodegradability. AI-driven lipid conversion processes enable the creation of **biodegradable plastics** from renewable resources, offering a sustainable alternative to petroleum-based plastics.
- **Waste-to-Value Solutions**: AI can optimize processes that convert organic waste into valuable bio-based chemicals, including lipids. This waste-to-value model reduces the burden on landfills and minimizes environmental pollution.

Example: In biofuel production, AI can optimize the conversion of algae lipids into biodiesel, significantly improving yield and reducing the energy intensity of the process. AI models can predict the best environmental conditions for algae growth, ensuring maximum efficiency in lipid production.

AI-Driven Carbon Capture and Utilization (CCU)

Another critical area where AI is making a significant impact is **carbon capture and utilization (CCU)**. As the world grapples with the consequences of high CO_2 emissions, innovative AI-powered solutions are emerging to capture and repurpose carbon, thus mitigating its environmental impact.

- **Direct Air Capture (DAC)**: AI algorithms can enhance the efficiency of DAC technologies by optimizing the capture of CO_2 directly from the atmosphere. This captured carbon can then be used in the production of biofuels, building materials, or other useful chemicals, contributing to a circular carbon economy.
- **Biological Carbon Capture**: AI models are being applied to biological systems, such as algae or plants, to enhance natural carbon sequestration processes. By optimizing these processes, AI can help increase the amount of CO_2 that is absorbed and stored in biomass, soils, or oceans.
- **Carbon-Neutral Production Systems**: AI-driven systems can be designed to integrate CCU technologies into industrial production processes, ensuring that the carbon released during manufacturing is captured and reused, leading to carbon-neutral or even carbon-negative production systems.

Example: AI has been used in optimizing microbial carbon fixation processes, where bacteria are engineered to absorb CO_2 and convert it into useful bio-based products like plastics or fuels. AI accelerates the research and scaling of these systems, offering a viable solution for large-scale carbon capture.

AI's Role in Promoting Economic and Social Sustainability

While AI-driven biochemical solutions can address environmental challenges, they also contribute to **economic and social sustainability**. By optimizing resource utilization, reducing waste, and promoting circular economies, AI is fostering more resilient and equitable systems across industries.

- **Green Economy and Job Creation**: The development and implementation of AI-powered environmental solutions can create new industries, including clean energy, sustainable agriculture, and waste management. These industries, in turn, generate jobs and drive economic growth, particularly in regions that are transitioning away from fossil fuel dependence.
- **Global Health Benefits**: AI-driven solutions for sustainable food production, carbon capture, and energy generation can have a positive impact on global health by reducing environmental pollution, improving food security, and mitigating the effects of climate change. These advancements improve the quality of life for communities worldwide.
- **Accessibility and Equity**: AI has the potential to make sustainable technologies more accessible to developing nations, helping them leapfrog traditional, polluting industries and adopt cleaner, more efficient solutions. By lowering the cost of energy and essential resources, AI can contribute to poverty alleviation and improved living standards.

Example: AI-powered systems that optimize energy use in agriculture can help small-scale farmers in developing regions increase crop yields while minimizing water and fertilizer consumption, leading to improved food security and income generation.

The Global Impact of AI-Driven Biochemical Solutions

AI-driven biochemical solutions are poised to have a profound impact on **global sustainability goals**. These technologies support a broad range of initiatives, including:

- **The United Nations Sustainable Development Goals (SDGs)**: AI-driven solutions contribute directly to SDGs such as clean energy, climate action, sustainable cities, and responsible consumption and production.
- **Climate Change Mitigation**: By enhancing carbon capture technologies and promoting sustainable energy production, AI helps combat climate change, reducing greenhouse gas emissions and promoting long-term planetary health.
- **Food and Water Security**: AI technologies in sustainable agriculture, water management, and bio-based product production support efforts to ensure that food and water resources are available for the growing global population.

Conclusion

AI-driven biochemical solutions represent a transformative force in addressing the world's most pressing environmental, social, and economic challenges. By enhancing lipid conversion processes and optimizing carbon capture and utilization, AI is playing a central role in shaping a sustainable future. As these technologies continue to evolve, they will drive the transition to a more sustainable and equitable global economy, ensuring that future generations inherit a healthier planet.

In the final chapter, we will reflect on the role of AI in shaping the future of biochemical solutions, focusing on how these technologies will evolve and contribute to the global pursuit of sustainability and human well-being.

Chapter 23: Challenges in AI-Driven Lipid Conversion and Carbon Capture

Introduction

Artificial Intelligence (AI) is transforming the way industries tackle some of the most pressing global challenges, from climate change to health disparities. In the field of biochemical solutions, AI has demonstrated its immense potential in areas such as lipid conversion for biofuel production and carbon capture and utilization (CCU). While the progress made with AI-driven technologies in these areas is significant, there remain several challenges that must be overcome to fully realize the potential of these innovations on a global scale.

This chapter explores the technical challenges in scaling up AI-powered lipid conversion processes and carbon capture systems, the data and resource limitations that hinder broader implementation, and the barriers to the widespread adoption of AI technologies in the biochemical industry. Additionally, we will examine strategies to address these challenges and make AI-driven biochemical solutions more accessible, efficient, and impactful for industries and society at large.

Technical Challenges in Scaling Up AI-Powered Lipid Conversion Processes

One of the key technical challenges in scaling up AI-driven lipid conversion processes lies in the **integration** of AI models with existing biochemical systems at an industrial scale. While AI can optimize lipid conversion at the laboratory level, translating these models to large-scale operations requires overcoming several barriers:

1. **System Complexity**: Biochemical systems are inherently complex, and large-scale processes involve numerous variables—temperature, pH, pressure, and concentration—that interact in ways that can be difficult to model accurately. AI models must be robust enough to account for this complexity without overfitting to the data from smaller-scale experiments.
2. **Real-Time Monitoring**: For AI to optimize lipid conversion in real-time, it requires continuous monitoring of the biochemical processes. However, scaling the technologies for real-time data collection and analysis presents technical difficulties, particularly in terms of sensor accuracy, data integration, and the speed of analysis.
3. **Optimization of Process Variables**: Industrial-scale lipid conversion often involves diverse feedstocks with varying lipid content. AI needs to constantly adjust process parameters to optimize yield and efficiency, and this requires advanced algorithms that can adapt to fluctuations in the feedstock and environmental conditions.

4. **Economic Feasibility**: The initial costs for implementing AI-driven lipid conversion processes at an industrial scale are high. These costs include not only the development and deployment of the AI models but also the infrastructure for data collection, storage, and processing. Ensuring economic feasibility remains a major challenge for many companies considering AI in their biochemical operations.

Data and Resource Limitations

AI technologies rely on large, high-quality datasets to generate accurate models. However, the availability of comprehensive data for training AI models in the biochemical field is often limited:

1. **Data Availability**: Obtaining sufficient data to train AI models is challenging in many biochemical applications. In areas like lipid conversion and carbon capture, datasets may be sparse, inconsistent, or incomplete, making it difficult for AI systems to generate reliable predictions and optimizations. For example, varying feedstocks and environmental conditions may yield inconsistent results, and a lack of standardization in data collection practices can further complicate the analysis.

2. **Data Quality**: AI models are only as good as the data they are trained on. Poor-quality or biased data can lead to inaccurate models, resulting in suboptimal process optimization or even negative environmental impacts. The problem is particularly prevalent in biochemical processes, where environmental variables such as temperature, humidity, and substrate variations can introduce noise into data collection.

3. **Resource Constraints**: Implementing AI-driven solutions requires considerable computing power and data storage capabilities. Biochemical companies, particularly smaller enterprises and those in developing regions, may struggle to meet the resource demands required for AI technologies. The cost of computing infrastructure and cloud-based storage can be prohibitive, especially when working with vast datasets.

4. **Lack of Standardized Data Protocols**: In the biochemical industry, there is often a lack of standardized data protocols, which makes it difficult to share and compare datasets across different research institutions or industrial settings. Standardization would facilitate the development of more reliable AI models by ensuring data consistency.

Overcoming Barriers to Widespread Adoption of AI in Biochemical Industries

Beyond the technical challenges, the adoption of AI-driven solutions in the biochemical industry faces several other barriers:

1. **Industry Resistance to Change**: Many industries, particularly in the traditional biochemical sectors, are resistant to change due to established systems and practices. The initial investment, uncertainty around ROI, and fear of job displacement may hinder the adoption of AI technologies. Organizations may also lack the necessary expertise to implement AI-driven solutions effectively.

2. **Regulatory and Compliance Issues**: The implementation of AI in the biochemical industry must comply with strict regulatory standards, particularly in areas such as pharmaceuticals, food production, and environmental safety. AI models must be transparent, explainable, and auditable to meet regulatory requirements. In some cases, regulatory bodies may not yet have established clear guidelines for the use of AI in these applications, creating uncertainty for companies looking to adopt these technologies.

3. **Ethical Concerns**: The use of AI in biochemical processes raises ethical concerns, particularly around data privacy, bias in AI models, and the equitable distribution of the benefits of AI. These concerns may discourage adoption, particularly in regions with strong ethical guidelines or where public trust in AI is limited.

4. **Skills Gap**: AI-driven biochemical solutions require highly skilled personnel, including data scientists, biochemists, and engineers. However, there is currently a shortage of professionals with the interdisciplinary knowledge required to implement AI solutions effectively in the biochemical sector. Companies must invest in training and education to bridge this skills gap.

Strategies to Address Challenges

Despite these challenges, several strategies can be employed to overcome barriers and enable the broader adoption of AI in the biochemical industry:

1. **Collaborations and Partnerships**: Industry collaborations between academia, research institutions, and biotech companies can help pool resources, data, and expertise to accelerate AI-driven innovation. Collaborative efforts can also provide smaller companies with access to the necessary infrastructure and expertise to deploy AI solutions.
2. **Public-Private Investment**: Governments and private investors can play a key role in supporting the adoption of AI by funding research, providing grants for pilot projects, and offering tax incentives for companies that implement AI technologies. Investments in AI-driven carbon capture technologies and sustainable biochemical solutions can provide long-term economic and environmental benefits.
3. **Open-Source Data and Model Sharing**: Encouraging the creation of open-source data repositories and sharing AI models across the industry can help address the data scarcity problem. Industry-wide data-sharing platforms would provide AI researchers with the vast datasets required for training and improve the generalizability of AI models.

4. **Standardization of Data Collection**: Industry groups can work together to develop standardized data collection protocols, ensuring consistency and quality in the data used to train AI models. Standardization will also make it easier to benchmark the performance of AI-driven solutions across different industrial settings.

5. **Regulatory Frameworks**: Governments and regulatory bodies can establish clear, transparent guidelines for the use of AI in biochemical industries. These frameworks should focus on safety, ethics, and accountability, ensuring that AI technologies are deployed responsibly and that industries comply with environmental and public health standards.

Conclusion

While AI-driven lipid conversion and carbon capture technologies hold immense promise for addressing global challenges such as climate change and food security, technical and organizational barriers remain significant. Overcoming these challenges will require a concerted effort from governments, industries, and academia to advance AI innovation, improve data availability and quality, and develop scalable, sustainable solutions. By addressing these barriers head-on, we can accelerate the deployment of AI-powered technologies that will help create a more sustainable and equitable future for all.

Chapter 24: The Future of AI in Biochemical Solutions for Human Health

Introduction

The potential for Artificial Intelligence (AI) to shape the future of biochemical solutions is vast, and its transformative role in improving human health and addressing global sustainability challenges is becoming increasingly apparent. As AI continues to evolve, its integration with biochemical processes such as lipid conversion, carbon capture, and metabolic optimization holds the promise of revolutionizing healthcare and environmental solutions alike.

In this chapter, we explore the future applications of AI in biochemical engineering, with a focus on how these technologies will continue to influence human health, disease prevention, and sustainable practices. We will also discuss how AI's ongoing advancements will shape the healthcare industry, the future of AI-driven biochemical solutions, and the long-term vision for sustainable, human-centered biochemical processes.

The Role of AI in Future Biochemical Solutions

AI's potential to drive the development of new biochemical solutions rests on its ability to process vast datasets, uncover patterns, and predict outcomes in ways that humans cannot. In the future, AI will be integral in optimizing and advancing several key areas of biochemical engineering, including:

Precision Medicine and Personalized Health

- **Tailored Drug Discovery**: AI will facilitate the discovery of new drugs by predicting how specific compounds will interact with the human body, particularly in lipid metabolism and other complex biological processes. By analyzing genetic, environmental, and lifestyle data, AI can help create highly personalized treatment plans that address individual patients' needs.
- **Optimized Treatment Pathways**: AI can analyze data from multiple sources (e.g., medical records, lab results, genetic data) to predict the most effective treatment pathways for diseases related to lipid metabolism, such as obesity, diabetes, and cardiovascular diseases. This allows for precision medicine that minimizes side effects and improves patient outcomes.

AI-Optimized Metabolic Engineering

- **Re-engineering Human Metabolic Pathways**: AI can play a crucial role in designing and optimizing metabolic pathways to improve lipid conversion processes in human cells. By simulating various biochemical reactions and predicting their outcomes, AI can help bioengineers create more efficient metabolic pathways that promote health and treat diseases.
- **AI-Assisted Gene Editing**: AI's ability to analyze vast genomic data can enhance the precision of CRISPR and other gene-editing technologies. By identifying specific genes responsible for metabolic dysfunctions, AI will guide the development of targeted genetic therapies that can correct or enhance lipid-related processes in humans.

Sustainable Solutions for Environmental Health

- **Carbon Capture and Utilization (CCU)**: AI will continue to refine carbon capture technologies by optimizing the processes for capturing and converting CO_2 into useful materials like biofuels or biodegradable plastics. AI models will predict the most efficient carbon capture systems and help design scalable solutions to reduce the environmental footprint of industrial activities.
- **Waste-to-Value Applications**: AI's role in waste-to-value processes will be expanded, as it helps convert waste products into useful bio-based chemicals, renewable energy, and sustainable materials. For example, waste carbon can be used to produce biofuels, and organic waste can be converted into biodegradable plastics, reducing landfill dependence and pollution.

AI-Driven Revolution in Healthcare

The healthcare sector stands to benefit immensely from the continued integration of AI into biochemical processes. Over the next decade, we will witness a significant transformation in the way diseases are diagnosed, treated, and prevented, thanks to AI's capacity to harness complex biological data. Here are some key areas where AI will play a pivotal role:

Early Detection and Diagnosis

- **Predictive Modeling for Disease Prevention**: AI will develop advanced models for early disease detection based on vast amounts of health data, including genetic information, lifestyle habits, and environmental factors. These AI models will be able to predict an individual's risk for conditions related to lipid metabolism (e.g., cardiovascular diseases) and offer preventive measures that can reduce the likelihood of developing these diseases.
- **AI-Enhanced Imaging and Diagnostics**: AI-powered diagnostic tools, such as machine learning-based imaging systems, will be able to identify abnormalities in tissues or organs related to metabolic disorders at an early stage. For example, AI can analyze scans to detect fatty liver disease or predict the likelihood of metabolic syndrome, allowing for faster interventions and treatments.

Clinical Decision Support Systems (CDSS)

- **AI for Personalized Treatment Plans**: By integrating AI into clinical decision support systems, healthcare providers will be able to offer more accurate and personalized treatment plans. AI systems will analyze patient data, including genetics, medical history, and lifestyle factors, to recommend tailored therapies and monitor their effectiveness in real-time.
- **Optimized Drug Dosing**: AI will also assist in optimizing drug dosing, especially for personalized medicine. Through real-time monitoring and continuous feedback loops, AI can ensure that patients are receiving the correct dosage of medications, minimizing the risk of side effects and improving therapeutic efficacy.

AI-Powered Drug Development

- **Accelerating Drug Discovery**: AI's capacity to rapidly analyze chemical structures, biological targets, and clinical data will streamline the process of drug discovery. With AI, pharmaceutical companies will be able to design drugs targeting specific metabolic pathways, including those involved in lipid conversion and energy production, allowing for faster and more efficient development of novel therapeutics.
- **Repurposing Existing Drugs**: AI will also accelerate the identification of existing drugs that may be repurposed for new applications. By analyzing vast datasets from clinical trials, AI can reveal new uses for approved drugs, offering solutions for diseases related to lipid metabolism, such as dyslipidemia, obesity, and type 2 diabetes.

The Future of AI in Sustainable Biochemical Solutions

In addition to its applications in human health, AI will continue to drive innovations in sustainable biochemical solutions. The future will witness the development of scalable and cost-effective AI-driven technologies that address both environmental and health-related challenges. Some key areas to watch for in the coming years include:

Green Chemistry and Environmental Sustainability

- **AI-Enhanced Green Chemistry**: AI will optimize green chemistry processes, enabling the production of bio-based chemicals and materials from renewable resources. This will reduce dependence on fossil fuels and minimize the environmental impact of chemical production processes.
- **Sustainable Agriculture**: AI will play an integral role in optimizing agricultural practices by helping to identify the most efficient ways to grow crops that are rich in lipids and other important compounds used in biochemical applications. Through AI-based precision farming techniques, we will be able to reduce resource use and increase crop yield, all while minimizing environmental degradation.

Waste Reduction and Circular Economy

- **AI-Driven Waste Management**: AI systems will revolutionize waste management by optimizing recycling processes and converting waste into valuable resources. From organic waste to plastics, AI will help identify the most efficient methods for repurposing waste materials, contributing to a circular economy.
- **AI for Sustainable Manufacturing**: AI will enhance the efficiency of manufacturing processes, reducing waste and energy consumption. By predicting demand and optimizing supply chains, AI will enable more sustainable production practices across industries, from biofuels to pharmaceuticals.

Long-Term Vision for AI in Biochemical Solutions

The long-term vision for AI in biochemical solutions is one where AI-powered systems operate autonomously and continuously to enhance human health and sustainability. These AI-driven systems will optimize metabolic processes in real time, detect and treat diseases before symptoms appear, and contribute to a more sustainable world by reducing waste and carbon emissions.

By harnessing the power of AI, we can look forward to a future where biochemical solutions are not only efficient and sustainable but also accessible to all, transforming how we approach healthcare, environmental protection, and resource management.

Conclusion

As we move forward into an increasingly AI-driven world, the potential for these technologies to reshape human health and environmental sustainability is immense. From personalized medicine and precision drug development to AI-powered carbon capture and sustainable biochemical production, the possibilities are endless. The future of AI in biochemical engineering holds the promise of better health outcomes, a cleaner environment, and a more sustainable and equitable future for generations to come. As we continue to advance in these areas, it is imperative that we do so responsibly, ensuring that AI is deployed ethically, equitably, and in alignment with the greater good.

Chapter 25: Conclusion: Leveraging AI for Ethical and Sustainable Biochemical Solutions

Introduction

As we reach the conclusion of this exploration into AI-driven biochemical solutions, it's clear that the marriage of artificial intelligence and biochemistry holds unprecedented potential for addressing some of the world's most pressing challenges. From revolutionizing human health to fostering environmental sustainability, AI's role in optimizing biochemical processes, particularly through lipid conversion and carbon capture, will continue to reshape industries, enhance healthcare, and promote a more sustainable future.

This final chapter summarizes the key insights from the book, reflecting on the progress we've made and the transformative impact that AI-driven innovations could have in the years to come. With these advancements, humanity stands at the cusp of a new era of scientific and industrial growth, grounded in responsible, ethical practices.

Key Insights from the Book

AI's Transformative Role in Biochemical Engineering

AI technologies, including machine learning, neural networks, and predictive modeling, have already begun to enhance our understanding of complex biochemical processes. From optimizing lipid metabolism in humans to improving carbon capture and utilization (CCU) in both biological and industrial settings, AI's impact is profound. The ability to model, simulate, and optimize biochemical pathways enables more efficient, scalable, and sustainable solutions that were previously out of reach.

AI-Driven Lipid Conversion and Human Health

Lipids are crucial in human metabolism, influencing everything from energy production to hormone regulation. Through AI-driven optimization, we can unlock the full potential of lipid conversion to improve treatments for metabolic disorders, such as obesity, diabetes, and cardiovascular diseases. AI not only assists in developing personalized treatment plans but also aids in identifying novel therapeutic pathways to address lipid-related health issues at a molecular level.

AI-Optimized Carbon Capture and Environmental Sustainability

One of the most promising applications of AI in the context of environmental sustainability is its role in optimizing carbon capture systems. By utilizing AI to design more efficient carbon capture technologies, we can significantly reduce greenhouse gas emissions and combat climate change. Furthermore, AI-powered solutions for waste-to-value processes—transforming waste into biofuels, biodegradable plastics, and other useful materials—offer an effective strategy for reducing pollution and advancing a circular economy.

Sustainability and Industrial Applications

AI's potential extends to industrial settings, where it can optimize large-scale biochemical production processes. Whether it's in the production of sustainable biofuels, biodegradable plastics, or pharmaceuticals, AI's ability to enhance process control and reduce resource waste is pivotal. AI can help scale these technologies in a way that makes them both economically viable and environmentally friendly, driving the transition to a greener, more sustainable industrial ecosystem.

Ethical and Responsible AI Use

As AI continues to influence biochemical engineering, it is essential to establish strong ethical guidelines for its application. This includes ensuring transparency, data privacy, and fairness, especially when it comes to personal health data used in personalized medicine. The responsible use of AI is crucial to avoid unintended consequences, such as biases in data or algorithmic decision-making. In a rapidly evolving field, ethical considerations must be at the forefront of AI development and deployment to ensure that its benefits are distributed equitably.

AI in Personalized Medicine

AI's ability to tailor medical treatments to individual patients is perhaps one of the most significant contributions it can make to human health. Through personalized medicine, AI can help optimize lipid metabolism and metabolic processes in unique ways for each patient. This could lead to faster recovery times, fewer side effects, and better overall outcomes, particularly for those suffering from chronic metabolic disorders.

Global Implications for Social and Economic Sustainability

On a global scale, AI-driven biochemical solutions have the potential to address issues related to food security, climate change, and economic inequality. AI can help optimize agricultural practices to improve crop yields, create more efficient food distribution networks, and ensure equitable access to healthcare and clean energy. Moreover, AI's impact on reducing carbon footprints in industries and urban environments will help advance global efforts to meet sustainability goals, such as the United Nations' Sustainable Development Goals (SDGs).

The Long-Term Vision for AI in Biochemical Engineering

The future of AI in biochemical engineering is a rapidly evolving frontier, and its long-term potential extends far beyond the current applications we've discussed. As AI continues to advance, we can anticipate even more innovative uses in healthcare, environmental sustainability, and industrial optimization. Some of the most exciting prospects include:

AI-Powered Systems for Real-Time Health Monitoring

In the future, AI will enable continuous, real-time monitoring of individual health metrics, offering personalized feedback to optimize metabolism and prevent disease onset. AI-powered wearable devices could provide constant updates on metabolic health, giving individuals and healthcare providers actionable insights to adjust treatment plans dynamically.

AI-Enhanced Regenerative Medicine

AI could revolutionize regenerative medicine by improving the precision of stem cell therapy and tissue engineering. Through AI's ability to model tissue growth and regeneration, we could see faster and more successful regenerative treatments for a variety of conditions, including those related to lipid metabolism, organ failure, and chronic diseases.

AI-Driven Circular Economy Models

In line with sustainability goals, AI will help build closed-loop systems within industries, where waste is minimized, and resources are continually reused. AI could manage energy consumption, waste production, and material sourcing to create more efficient processes that promote a circular economy.

AI in Global Health Crisis Management

In the face of future pandemics or global health emergencies, AI's predictive capabilities will allow for rapid response and management. AI models could simulate disease spread, predict medical resource needs, and optimize vaccination or treatment strategies in real-time, ultimately saving lives and reducing the global impact of health crises.

Concluding Thoughts: Responsible Use of AI in Biochemical Engineering

The potential of AI to drive ethical and sustainable biochemical solutions cannot be overstated. However, as with any powerful technology, it is essential that we use AI responsibly, keeping in mind the broader societal, environmental, and ethical implications. Ensuring that AI is developed and deployed with a focus on human well-being, equity, and sustainability is essential for maximizing its potential and minimizing harm.

As we look to the future, the integration of AI with biochemical processes holds the promise of a healthier, more sustainable world. By continuing to foster innovation, encouraging collaboration across industries, and developing robust ethical frameworks, we can harness the full potential of AI to create a future where both people and the planet thrive.

Final Thoughts

The journey of AI-driven biochemical engineering is just beginning. What we have outlined in this book represents only the first steps toward a future where AI is seamlessly integrated into every aspect of our biochemical processes. The transformative potential of AI in enhancing human health and sustainability is limitless, and with the right guidance and ethical oversight, AI can be a powerful force for good in the world. As we look to the horizon, we must remain committed to leveraging these innovations responsibly, ensuring that the solutions we create are not only effective but also just, equitable, and sustainable.

www.ingramcontent.com/pod-product-compliance
Lightning Source LLC
Chambersburg PA
CBHW082245220526
45469CB00009B/2881